EPA 832-B-03-001
March 2003

Voluntary National Guidelines for Management of Onsite and Clustered (Decentralized) Wastewater Treatment Systems

Office of Water
Office of Research and Development
U.S. Environmental Protection Agency

CONTENTS

EXECUTIVE SUMMARY

The performance of onsite and clustered (decentralized) wastewater treatment systems is a national issue of great concern to the Environmental Protection Agency (EPA). Decentralized systems are used in 25 percent of the homes in the United States and 33 percent of new development, and they are permanent components of our nation's wastewater infrastructure. Decentralized wastewater treatment systems are defined here as managed individual onsite or clustered wastewater systems (commonly referred to as septic systems, private sewage systems, individual sewage treatment systems, onsite sewage disposal systems, or "package" plants) used to collect, treat, and disperse or reclaim wastewater from individual dwellings, businesses, or small communities or service areas. Unfortunately, many of the systems in use are improperly managed and do not provide the level of treatment necessary to adequately protect public health and surface and ground water quality. Proper management of decentralized systems involves implementation of a comprehensive, life-cycle series of elements and activities that address

public education and participation, planning, performance, site evaluation, design, construction, operation and maintenance, residuals management, training and certification/licensing, inspections and monitoring, corrective actions, recordkeeping/inventorying/reporting, and financial assistance and funding.

Therefore, EPA is issuing *Voluntary National Guidelines for Management of Onsite and Clustered (Decentralized) Wastewater Treatment Systems* (referred to

as the Management Guidelines) to enhance the performance and reliability of decentralized wastewater treatment systems through improved

> *Decentralized systems are used in 25% of U.S. homes and are permanent components of our nation's wastewater infrastructure.*

management programs. The Management Guidelines will help improve system performance by encouraging institutionalizing the concept of management; raising the quality of state, tribal, and local management programs; and suggesting minimum levels of activity. Adequately managed decentralized systems that protect the environment and public health can provide an alternative to centralized wastewater treatment systems. EPA continues to support the most sustainable approach to implementing protective water pollution control solutions whether it be centralized or decentralized. The Management Guidelines are intended to be used when a decision to implement a decentralized approach is or has been made. They complement any other applicable federal, state, tribal, or local government requirements, including the National Pollutant Discharge Elimination System (NPDES) program under the Clean Water Act (CWA) and the Underground Injection Control (UIC) program under the Safe Drinking Water Act (SDWA).

EPA intends that state, tribal, and local authorities use the Management Guidelines, along with other applicable federal requirements, to help communities in meeting water quality and public health goals. The Management Guidelines can be integrated into a comprehensive watershed approach at the state, tribal, or local government level. The benefits of an adequate management program include protection of water quality and public health, protection of consumers' investment in home and business ownership, increased onsite system service life and replacement cost savings, avoidance of transfers of water away from the source by conserving ground water, and elimination of the need to use a community's tax base to finance sewers. As noted above, and in more detail later in this document, proper management is a comprehensive term for achieving the long-term sustainability of a system, including adequate operation and maintenance of the system. Although implementation of the Management Guidelines is voluntary, EPA strongly encourages considering them as a template in strengthening existing management programs and implementing new ones.

Not in My Septic System!

X **Cloggers**
diapers, cat litter, cigarette filters, coffee grounds, grease, feminine hygiene products, etc.

X **Killers**
household chemicals, gasoline, oil, pesticides, antifreeze, paint, etc.

Unfortunately, although some management programs are effective, many existing state, tribal, and local rules that regulate onsite systems are not adequate to ensure proper performance. "Failure" of onsite systems is a term subject to much debate; however, 1995 U.S. Census data report that over 10 percent of all systems back up into homes or have wastewater emerging on the ground surface, and that more than half the systems in the United States were installed more than 30 years ago when onsite rules were nonexistent or poorly enforced. Few systems receive proper maintenance because homeowners are either unaware of the need for maintenance or find it a distasteful task. In addition, most regulatory programs do not require homeowner accountability for system performance after installation. Although it is difficult to measure and document specific cause-and-effect relationships between onsite wastewater treatment systems and the quality of our water resources, it is widely accepted that improperly managed systems contribute to major water quality problems. The *National Water Quality Inventory 1996 Report to Congress* states that "improperly constructed and poorly maintained septic systems are believed to cause substantial and widespread nutrient and microbial contamination to ground water." Ultimately it is the absence of a comprehensive management program addressing each of these issues that prevents onsite and clustered (decentralized) systems from being considered as an effective and reliable wastewater treatment strategy. Consequently, the potential for health and water quality problems from poorly managed systems is increasing.

If effectively implemented by state, tribal, and local governments, the Management Guidelines might provide for a viable, long-term option for meeting public health and water quality goals, particularly for small and rural communities. In addition, appropriate management programs will support the activities and approaches being used in other EPA programs and contribute toward achievement of mutual water quality and public health goals. These programs include Watershed Management, National Pollutant Discharge Elimination System, Biosolids and Residuals Management, Storm Water Management, Water Quality Management (including Total Maximum Daily Loads, or TMDLs), Water Quality Standards, Source Water Assessment and Protection, Underground Injection Control, Coastal Zone Management, Nonpoint Source Control Program, and Technology Transfer.

In deciding whether to use onsite systems, it is important to consider the risks they might pose to the

Few systems receive proper maintenance... most regulatory programs do not require homeowner accountability for system performance.

environment and public health. There may be cases where onsite systems are not appropriate because of the environmental sensitivity or public health concerns of an area. In the cases where onsite systems are appropriate, it is critical that they be managed to prevent environmental and public health impacts.

Five management models are provided as conceptual approaches with progressively increasing management controls as sensitivity of the environment and/or treatment system complexity increases (see box below). Each model consists of 13 critical elements that describe activities to be performed to achieve the management goal. The purpose of the models is to provide a guide to match the needed management controls to the potential public health and water quality risks presented by decentralized systems in a particular area. The models are flexible so that programs can be customized by substituting elements of one program into another to accommodate local needs, practices, and conditions. The models are built around ensuring the accountability and competency of regulators and service providers through certification and continuing education, owners through education and/or inspection requirements, and third-party managers through contract and permit stipulations to achieve their goals. The "best" model program for a community is not necessarily in the higher levels, but rather is the model that provides the most appropriate management controls for the potential risks.

The Five Management Models

- **Management Model 1 - "Homeowner Awareness"** specifies appropriate program elements and activities where treatment systems are owned and operated by individual property owners in areas of low environmental sensitivity. This program is adequate where treatment technologies are limited to conventional systems that require little owner attention. To help ensure that timely maintenance is performed, the regulatory authority mails maintenance reminders to owners at appropriate intervals.

- **Management Model 2 - "Maintenance Contracts"** specifies program elements and activities where more complex designs are employed to enhance the capacity of conventional systems to accept and treat wastewater. Because of treatment complexity, contracts with qualified technicians are needed to ensure proper and timely maintenance.

- **Management Model 3 - "Operating Permits"** specifies program elements and activities where sustained performance of treatment systems is critical to protect public health and water quality. Limited-term operating permits are issued to the owner and are renewable for another term if the owner demonstrates that the system is in compliance with the terms and conditions of the permit. Performance-based designs may be incorporated into programs with management controls at this level.

- **Management Model 4 - "Responsible Management Entity (RME) Operation and Maintenance"** specifies program elements and activities where frequent and highly reliable operation and maintenance of decentralized systems is required to ensure water resource protection in sensitive environments. Under this model, the operating permit is issued to an RME instead of the property owner to provide the needed assurance that the appropriate maintenance is performed.

- **Management Model 5 - "RME Ownership"** specifies that program elements and activities for treatment systems are owned, operated, and maintained by the RME, which removes the property owner from responsibility for the system. This program is analogous to central sewerage and provides the greatest assurance of system performance in the most sensitive of environments.

The legal authority for regulating onsite and clustered (decentralized) wastewater treatment systems generally rests with state, tribal, and local governments. EPA recognizes that these units of government need a flexible framework and guidance to tailor their programs to the specific needs of communities and watersheds. Although each management program model stands alone, the models are intended only to be guides in developing an appropriate management program. Activities in program elements of higher-level models may be incorporated into lower-level programs to assist the local program in achieving its desired objectives. Also, it is possible to implement more than one management program model within a jurisdiction as appropriate for the circumstances encountered, such as housing density, receiving environment characteristics, new development, high-volume or high-strength wastewaters, and so forth. Management models may also be implemented in conjunction with centralized wastewater treatment and collection. It is important to note that the management program models are not intended to supersede existing federal, state, tribal, or local laws and regulations, but rather to complement them in protecting public health and water quality.

> It is important to note that the management program models are not intended to supersede existing federal, state, tribal, or local laws and regulations.

To assist state, tribal, and local units of government in evaluating and upgrading their onsite and clustered (decentralized) wastewater management programs, a draft *Handbook for Management of Onsite and Clustered (Decentralized) Wastewater Treatment Systems* (referred to as the Management Handbook) complements the Management Guidelines. The draft Management Handbook includes case studies and examples of materials used by communities that have implemented management programs effectively.

Substantial resources are available as well, including EPA's *Onsite Wastewater Treatment Systems Manual*, to assist regulatory agencies and communities in assessing the technical foundation of the elements and activities in their existing or considered management programs.

INTRODUCTION

What Is the Purpose of the Voluntary National Management Guidelines?

EPA has developed the *Voluntary National Guidelines for Management of Onsite and Clustered (Decentralized)* 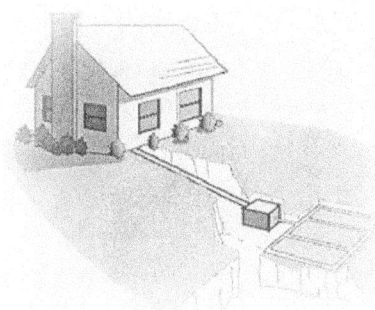 *Wastewater Treatment Systems* to raise the level of performance of onsite and clustered wastewater treatment systems through improved management programs. Decentralized wastewater treatment systems are defined here as individual onsite or clustered wastewater systems (commonly referred to as septic systems, private sewage systems, individual sewage treatment systems, onsite sewage disposal systems, or "package" plants) used to collect, treat, and disperse or reclaim wastewater from individual dwellings, businesses, or small communities and service areas. Such systems may provide an alternative to conventional centralized wastewater systems. However, any onsite or clustered wastewater treatment system that discharges pollutants from a point source to waters of the United States is subject to the National Pollutant Discharge Elimination System (NPDES) program under the Clean Water Act (CWA). Such discharge is illegal and subject to enforcement action unless it is authorized by an NPDES permit issued by an authorized state or tribe or by EPA. Onsite and clustered

EPA continues to support the most environmentally sound and cost-effective approach to implementing protective water pollution control solutions whether it be centralized or decentralized. The Management Guidelines are intended to be used when the decision is to implement a decentralized approach.

systems can be protective of public health and water quality if they are properly planned, sited, designed, constructed, installed, operated, and maintained. EPA is issuing this guidance to raise the quality of management programs, suggest minimum levels of activity, and encourage institutionalizing the concept of management. Implementation of the Management Guidelines can help communities meet water quality and public health goals, provide a greater range of options for cost-effectively meeting wastewater needs, and protect consumers' investment in home and business ownership. In a number of instances, decentralized wastewater treatment systems without proper management programs have failed in the long term because of lack of proper operation and/or maintenance and have had to be replaced by centralized systems. If centralized collection systems are feasible, decentralized systems are recommended only where there is assurance of an enforceable management system consistent with this strategy, including long-term financial and technical capacity for operation and maintenance.

These Management Guidelines are not intended to supercede any otherwise applicable federal, state, tribal, or local requirements. Also, the decision on use of centralized or decentralized wastewater treatment is one to be made at the state, tribal, or local level after consideration of a number of factors.

Please note that the statutes and regulations described in this document contain legally binding requirements. The guidance provided in this document does not substitute for those statutes or regulations. These Management

Guidelines are strictly voluntary and, by themselves, do not impose legally binding requirements on EPA, state, local, or tribal governments or members of the public and, based upon the circumstances, may not apply to a particular situation. Although EPA strongly recommends the approach outlined in this document, state and local decision makers are free to adopt approaches that differ from these Management Guidelines.

What Is Management?

Management of decentralized systems is implementation of a comprehensive, life-cycle series of elements and activities that address public education and participation, planning, performance, site evaluation, design, construction, operation and maintenance, residuals management, training and certification/licensing, inspections/monitoring, corrective actions,

recordkeeping/ inventorying/ reporting, and financial assistance and funding. Therefore, a management program involves, in varying degrees, regulatory and elected officials, developers and builders, soil and site evaluators, engineers and designers, contractors and installers, manufacturers, pumpers and haulers, inspectors, management entities, and property owners. Establishing the distinct roles and responsibilities of the partners involved is very important to ensuring proper system management.

Who Can Benefit from the Management Guidelines?

The Management Guidelines contain a set of management models, based on a comprehensive approach that relies on coordinating responsibilities and actions among the state, tribal, or local regulatory agency, the management entity or service provider,

and the system owner. EPA recognizes the importance of each party in improving management programs and encourages identification of distinct and separate roles and responsibilities when implementing management programs. The primary audiences for these Management

Guidelines are state, tribal, and local regulators and community officials who are responsible for regulating onsite and clustered systems. The secondary audiences include planners, designers, installers, operators, pumpers, haulers, management entities, and inspectors.

In particular, local communities with a need to improve system performance should consider these Management Guidelines as a first step in evaluating their existing programs. EPA also encourages state and tribal agencies that regulate onsite and clustered systems to evaluate their existing programs and address the program elements and activities detailed in these management models in their regulatory/management function. Although very important to implementation of a management program, owner responsibilities are not discussed here in detail. Materials to help owners improve management of their systems are provided in EPA's draft Management Handbook, which is being issued concurrently with these Management Guidelines.

To What Types of Systems Are the Guidelines Relevant?

The Management Guidelines are relevant to both existing communities and areas of new development that use onsite and clustered systems of any size for residential and commercial wastewater treatment and dispersal. Centralized collection and treatment facilities

are not addressed here. Industrial wastewater treatment systems are also not addressed because many industrial wastes are prohibited by federal and state regulation from using onsite treatment and dispersal, because of the potential to interfere with wastewater treatment, and/or to pollute ground water resources.

These Management Guidelines are not intended to be used to determine appropriate or inappropriate uses of land. The information in the Management Guidelines is intended to be used to help select appropriate management strategies and technologies that minimize risks to human health and water resources in areas where connections to centralized wastewater collection and treatment systems are not considered appropriate. The determination of appropriate siting requirements, system density restrictions, or required technologies is a state, tribal, or local decision. Substantial resources are available to ensure these decisions are sound; they are detailed in the draft Management Handbook.

What Are Management Guidelines?

These Management Guidelines consist of five models that are structured to reflect an increasing need for more comprehensive management as the sensitivity of the environment or the degree of technological complexity increases. A management program's intensity increases progressively from one management model to another, reflecting the increased level of management activities needed to achieve water quality and public health goals. A community would establish a management level

that is sufficient for its management needs. Although adoption of the Management Guidelines is voluntary, EPA strongly encourages communities to consider the Management Guidelines as a basis for their onsite and clustered wastewater management programs because of the public health and water quality concerns associated with these systems.

Why Are Management Guidelines Needed?

The performance of onsite and clustered wastewater treatment systems is a national issue of great concern to EPA. Onsite and clustered wastewater treatment systems serve approximately 25 percent of U.S. households (about 25 million) and approximately 33 percent of new development.[1] Onsite and clustered systems can provide a high level of public health and natural resource protection if they are properly planned, sited, designed, constructed, operated, and maintained.

> *More than half the existing onsite systems are over 30 years old, and surveys indicate at least 10 percent of these systems back up onto the ground surface or into the home each year.*

Unfortunately, many of the systems in use do not provide the level of treatment necessary to adequately protect public health or surface and ground water quality. Many were initially sited and installed as temporary solutions as a result of the perception that centralized treatment and collection would soon replace them. Comprehensive, life-cycle management did not play a role in the approval or the ongoing operation of many systems. More than half the existing onsite systems are over 30 years old, and surveys indicate at least 10 percent of these systems back up onto the ground surface or into the home each year.[1] Other data have shown that at least 20 percent of systems are malfunctioning to some degree.[2] In most cases the homeowner is not aware of a system failure until sewage backs up into the home or breaks out on the ground surface. In many places, local authorities lack records of many of the systems in the service area.

Although it is difficult to measure and document specific cause-and-effect relationships between onsite wastewater treatment systems and the quality of our water resources, it is widely accepted that improperly managed systems (resulting from inadequate siting, design, construction, installation, operation, and/or maintenance) contribute to major water quality problems. As documentation becomes available concerning the source of impairments, EPA will be better able to determine the extent of the relationship. It is already evident that improved operation and performance of onsite and clustered systems through better management practices will be essential if the nation's water quality and public health goals are to be attained.

In the *National Water Quality Inventory: 1996 Report to Congress,* state agencies designated the top 10 potential contaminant sources that threaten their ground water resources. The second most frequently cited contamination source was septic systems. The

report states that "improperly constructed and poorly maintained septic systems are believed to cause substantial and widespread nutrient and microbial contamination to ground water." Other contaminant sources identified by states included underground storage tanks, landfills, large industrial facilities, and numerous other activities.[3] States also identified more than 500 communities in the *1996 Clean Water Needs Survey*[4] as having failed septic

systems that have caused public health problems. In 1996 states reported septic systems as a leading source of pollution for more than one-third (36 percent) of the impaired miles of ocean shoreline surveyed.[3] Other leading sources included urban runoff and storm sewers, municipal sewer discharges, and industrial point sources. In U.S. classified shellfish growing areas, closures and harvest restrictions have occurred primarily because of "the concentration of fecal coliform bacteria associated with human sewage and with organic wastes from livestock and wildlife." The 1995 National Shellfish Register indicated that the most common pollution source cited for shellfish restrictions

> *The second most frequently cited contamination source (of ground water) is septic systems.*

was urban runoff (principal or contributing factor in 40 percent of all harvest-limited growing areas), followed by unidentified upstream sources (39 percent), wildlife (38 percent) and septic tanks (32 percent).[5] Onsite wastewater treatment systems might also be contributing to an overabundance of nutrients in ponds, lakes, and coastal estuaries, leading to overgrowth of algae and other nuisance aquatic plants. For example, the 45,000 septic systems in Sarasota County, Florida, contribute four times more nitrogen to Sarasota Bay than the City of Sarasota's advanced wastewater treatment plant.[6]

Onsite and clustered wastewater treatment systems also contribute to contamination of drinking water sources. EPA estimates that 168,000 viral illnesses and 34,000 bacterial illnesses occur each year as a result of consumption of drinking water from systems that rely on improperly treated ground water.[7] The contaminants of primary concern in EPA's study of ground water-based drinking water systems are waterborne pathogens from fecal contamination. Malfunctioning septic systems are identified as a potential source of this contamination;

other sources could include leaking or overflowing sanitary sewer lines, as well as storm water runoff. A recent example of contamination involved nearly 800  visitors to a fair in Washington County, New York, who became ill after consuming water from a well source that had likely been contaminated by a septic system at an adjacent dormitory. Other examples in which pollution was attributed to septic systems include 82 cases of shigellosis resulting from a contaminated well in Island Park, Idaho, in 1995; 46 cases of hepatitis A from a privately owned water supply in Racine, Missouri; and 49 cases of hepatitis A in Lancaster, Pennsylvania, in 1980.[8] EPA is concerned about the presence of nitrates in ground water, particularly in rural areas where residents must rely on individual wells and onsite systems to serve relatively small lots.

What Are the Benefits of a Management Program?

Benefits of a management program are accrued by both the communities developing effective management programs and the individual property owners. They include the following:

- *Protection of public health and local water resources.* Although unquantified, septic system failures in the form of yard backups have been recognized as a public health hazard and an insult to natural resources for many years. Improved management practices will minimize the occurrence of failures by ensuring (with proper planning, siting, design, installation, operation and maintenance, and monitoring) that pollutants are adequately treated and dispersed into the

environment, thereby reducing risks to public health and local water resources.

- *Protection of property values.* There are many documented instances over the past few decades of property values increasing in areas formerly served by failing onsite systems after the area has been sewered. Management programs offer an opportunity to obtain the same level of service and aesthetics as sewered communities at a fraction of the cost, thus providing property appreciation and cost savings.

- *Ground water conservation.* A well-managed onsite system will contribute to ground water recharge. Many areas of the United States that have undergone rapid development and sewering are experiencing rapidly declining water tables or water shortages because ground water is no longer being recharged by onsite systems.

- *Preservation of tax base.* A well-managed onsite system will prevent small communities from having to finance the high cost of centralized sewers. Many small communities have exhausted their tax base, at the expense of other public safety and education programs, to pay for those sewers. Many communities then entice growth in an effort to pay for the systems, thus destroying the community structure that originally attracted residents.

- *Life-cycle cost savings.* There is a clear indication that in many cases management may pay for itself in terms of lower failure rates and alleviation of the need for premature system replacement; however, this will depend on the types of systems employed and the management program chosen. Documentation of that savings is only now being initiated.

How Were the Management Guidelines Developed?

In April 1997 EPA published its *Response to Congress on Use of Decentralized Wastewater Treatment Systems,* which concluded that, overall, "adequately managed decentralized wastewater treatment systems are a

cost-effective and long-term option for meeting public health and water quality goals, particularly in less densely populated areas [small and rural communities]."[9] EPA stated that both centralized and decentralized system alternatives should be considered when upgrading failing onsite systems. The report found

> *Adequately managed decentralized wastewater treatment systems are a cost-effective and long-term option for meeting public health and water quality goals.*

that decentralized systems can protect public health and the environment, typically have lower capital and maintenance costs for low-density communities, are appropriate for varying site conditions, and are suitable for ecologically sensitive areas when adequately managed.

More important, EPA identified several major barriers to the increased use of these systems, including the lack of adequate management programs. Most onsite

and clustered systems are regulated at the state, tribal, or local level, not at the federal level, and there is a great deal of inconsistency in the regulatory approaches. Many existing management programs are inadequate or too narrow in focus, allowing premature system failures to occur. Although the varying reasons for system failure may include shortcomings in siting, design, construction, operation, or maintenance, it is ultimately the absence of a comprehensive management program—which addresses each of these issues—that prevents onsite and clustered systems from reaching their potential as an effective, reliable wastewater treatment strategy.

RELATIONSHIP TO OTHER WATER PROGRAMS

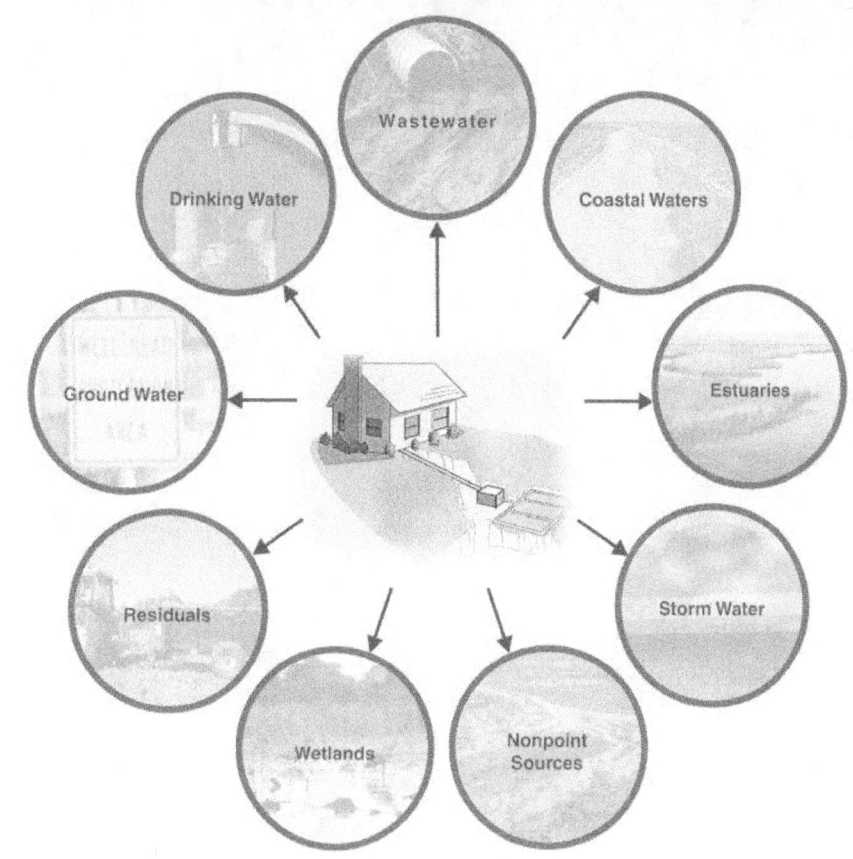

These Management Guidelines will help support the activities and approaches being applied in several other EPA programs and contribute toward achieving mutual water quality objectives and public health protection goals. The Management Guidelines complement any applicable regulatory authority under the Clean Water Act (CWA), Safe Drinking Water Act (SDWA), Coastal Zone Management Act/Coastal Zone Act Reauthorization Amendments of 1990 (CZMA/CZARA), or any other federal law. For example, there are certain situations where use of these Management Guidelines includes authorization under an NPDES permit, which is required for all discharges of pollutants from a point source to waters of the United States.

Related programs include, among others, Watershed Management, National Pollutant Discharge Elimination System, Biosolids and Residuals Management, Storm Water Management, Water Quality Management (including Total Maximum Daily Loads, or TMDLs), Water Quality Standards, Source Water Assessment and Protection, Underground Injection Control, Coastal Zone Management, Nonpoint Source Control Program, and Technology Transfer. The relationship of the Management Guidelines to these companion programs is summarized in Appendix B.

DESCRIPTION OF MANAGEMENT MODELS

Introduction

The Management Guidelines consist of a series of five management models. As the models progress from the Homeowner Awareness Model to the Responsible Management Entity (RME) Ownership Model, they reflect the need for improved management practices and increased oversight as determined by the complexity of

treatment systems employed and the potential risks to public health and water resources. For example, the Homeowner Awareness Model recommends management practices for areas where the risks to public health and water resources are low and the suitable treatment technologies are passive and robust. The RME Ownership Model, on the other hand, defines an appropriate level of practice and oversight for communities where there are significant risks to public health or water resources. Table 1, "Summary of Management Models," presents a brief description of each management model. The table presents the management program objectives, provides a brief description of the types of systems applicable, and lists the major benefits and limitations of each of the five management models.

Key Concepts

The Management Guidelines contain certain key concepts that are the foundation of changes needed to improve the performance of decentralized wastewater

Key Concepts

- **<u>An increase in the level of management as the level of risk and technical complexity increase</u>**

- **Inventorying** existing systems and their level of performance as a minimum

- **Operating permits** for large systems and clusters of onsite systems

- **Discharge permits** for systems that discharge to surface waters

- Increased requirements for **certification and licensing** of practitioners

- Elimination of **illicit discharges** to storm drains or sewers

treatment systems (see box above). These concepts are imbedded in the activities of each management model and have the potential to make a difference in the field.

Management Models

Tables 1 through 5 in Appendix A describe the management models, which include the objective or goal to be reached and an accompanying set of program elements and activities appropriate for achieving the stated objectives. The management models provide benchmarks for a state, tribal, or local unit of government to (1) select appropriate management objectives to meet its wastewater treatment needs, (2) evaluate the strengths and weaknesses of its current program in achieving the desired objectives, (3) design a management program and activities needed to meet unique local

Table 1: Summary of Management Models

TYPICAL APPLICATIONS	PROGRAM DESCRIPTION	BENEFITS	LIMITATIONS
MODEL 1 - HOMEOWNER AWARENESS MODEL			
• Areas of low environmental sensitivity where sites are suitable for conventional onsite systems.	• Systems properly sited and constructed based on prescribed criteria. • Owners made aware of maintenance needs through reminders. • Inventory of all systems	• Code-compliant system. • Ease of implementation; based on existing, prescriptive system design and site criteria. • Provides an inventory of systems that is useful in system tracking and area-wide planning.	• No compliance/problem identification mechanism. • Sites must meet siting requirements. • Cost to maintain database and owner education program.
MODEL 2 - MAINTENANCE CONTRACT MODEL			
• Areas of low to moderate environmental sensitivity where sites are marginally suitable for conventional onsite systems due to small lots, shallow soils, or low-permeability soils. • Small clustered systems.	• Systems properly sited and constructed. • More complex treatment options, including mechanical components or small clusters of homes. • Requires service contracts to be maintained. • Inventory of all systems. • Service contract tracking system.	• Reduces the risk of treatment system malfunctions. • Protects homeowner investment.	• Difficulty in tracking and enforcing compliance because it must rely on the owner or contractor to report a lapse in a valid contract for services. • No mechanism provided to assess effectiveness of maintenance program.
MODEL 3 - OPERATING PERMIT MODEL			
• Areas of moderate environmental sensitivity such as wellhead or source water protection zones, shellfish growing waters, or bathing/water contact recreation. • Systems treating high-strength wastes or large-capacity systems.	• Establishes system performance and monitoring requirements. • Allows engineered designs but may provide prescriptive designs for specific receiving environments. • Regulatory oversight by issuing renewable operating permits that may be revoked for noncompliance. • Inventory of all systems. • Tracking system for operating permit and compliance monitoring. • Minimum for large-capacity systems.	• Allows systems in more environmentally sensitive areas. • Operating permit requires regular compliance monitoring reports. • Identifies noncompliant systems and initiates corrective actions. • Decreases need for regulation of large systems. • Protects homeowner investment.	• Higher level of expertise and resources for regulatory authority to implement. • Requires permit tracking system. • Regulatory authority needs enforcement powers.
MODEL 4 - RESPONSIBLE MANAGEMENT ENTITY (RME) OPERATION AND MAINTENANCE MODEL			
• Areas of moderate to high environmental sensitivity where reliable and sustainable system operation and maintenance (O&M) is required, e.g., sole source aquifers, wellhead or source water protection zones, critical aquatic habitats, or outstanding value resource waters. • Clustered systems.	• Establishes system performance and monitoring requirements. • Professional O&M services through RME (either public or private). • Provides regulatory oversight by issuing operating or NPDES permits directly to the RME. (System ownership remains with the property owner.) • Inventory of all systems. • Tracking system for operating permit and compliance monitoring.	• O&M responsibility transferred from the system owner to a professional RME that is the holder of the operating permit. • Identifies problems needing attention before failures occur. • Allows use of onsite treatment in more environmentally sensitive areas or for treatment of high-strength wastes. • Can issue one permit for a group of systems. • Protects homeowner investment.	• Enabling legislation may be necessary to allow RME to hold operating permit for an individual system owner. • RME must have owner approval for repairs; may be conflict if performance problems are identified and not corrected. • Need for easement/right of entry. • Need for oversight of RME by regulatory authority.
MODEL 5 - RESPONSIBLE MANAGEMENT ENTITY (RME) OWNERSHIP MODEL			
• Areas of greatest environmental sensitivity where reliable management is required. Includes sole source aquifers, wellhead or source water protection zones, critical aquatic habitats, or outstanding value resource waters. • Preferred management program for clustered systems serving multiple properties under different ownership (e.g., subdivisions).	• Establishes system performance and monitoring requirements. • Professional management of all aspects of decentralized systems through public/private RMEs that own or manage individual systems. • Qualified, trained, owners and licensed professional owners/operators. • Provides regulatory oversight by issuing operating or NPDES permit. • Inventory of all systems. • Tracking system for operating permit and compliance monitoring.	• High level of oversight if system performance problems occur. • Simulates model of central sewerage, reducing the risk of noncompliance. • Allows use of onsite treatment in more environmentally sensitive areas. • Allows effective area-wide planning/watershed management. • Removes potential conflicts between the user and RME. • Greatest protection of environmental resources and owner investment.	• Enabling legislation and/or formation of special district may be required. • May require greater financial investment by RME for installation and/or purchase of existing systems or components. • Need for oversight of RME by regulatory authority. • Private RMEs may limit competition. • Homeowner associations may not have adequate authority.

Note: If applicable, NPDES requirements under the CWA or UIC requirements under the SDWA supercede any less stringent or inconsistent provision.

objectives, and (4) develop a plan for implementing the management program. The draft Management Handbook, which is being issued concurrently with these Management Guidelines, provides detailed guidance on how to select, evaluate, develop, and implement the Management Guidelines.

Evaluation of Risk

In deciding whether to use onsite systems, it is important to consider the risks they may pose to the environment and public health. There may be cases where onsite systems are not appropriate because of the environmental sensitivity or public health concerns of

an area. In the cases where onsite systems are appropriate, it is critical that they be managed to prevent environmental and public health impacts. All of the management models share the common goal of ensuring that public health and water resources are protected. Effective implementation of management programs requires coordination among state, tribal, and local water quality, public health, and planning and zoning agencies, and community officials. EPA continues to encourage this coordination on a watershed basis. Zoning ordinances and land use planning are also mechanisms that state, tribal, and local governments use to address water resource issues. Coordination is necessary as well to help ensure that state, tribal, and local decentralized wastewater programs are managed on a watershed basis to achieve protection consistent with applicable state and tribal water quality standards, including pathogen and nutrient

EPA recognizes that these units of government need a flexible framework and guidance to best tailor their management programs to the specific needs of the community and the needs of the watershed.

criteria. EPA believes that these goals are best achieved where performance-based management of onsite and clustered systems has been implemented to protect the quality of the receiving watershed and/or aquifer.

Flexibility Needed for Implementation

The legal authority for regulating onsite and clustered systems generally rests with state, tribal, and local governments. EPA recognizes that these units of

government need a flexible framework and guidance to best tailor their management programs to the specific needs of the community and the needs of the watershed. Although each management model stands alone, the models are intended only to be guides in developing an appropriate management program. Activities shown in program elements from one management model may be incorporated into another model to enhance the effectiveness of local programs in achieving the desired objectives under the prevailing circumstances. However, substituting activities from higher levels into lower-level management programs should be carefully considered because of the interdependence of many activities on overall program capabilities. It is also possible to implement more than one management model, as appropriate, within a jurisdiction for the circumstances encountered (housing density, site and soil characteristics, and treatment technology complexity). Further, it is important to note that these management models are not intended to supersede existing federal, state, tribal, and local laws and regulations, but rather to complement their role in protecting public health and water quality.

Roles and Responsibilities

Governmental roles and authority in implementation of management programs based on the Management Guidelines will vary from jurisdiction to jurisdiction. Application of the NPDES program under the CWA is required if there is a discharge of pollutants from a point source to a water of the United States. Similarly, application of the UIC program under the SDWA is

required if a large-capacity system is subject to UIC controls. The provisions of the program elements in each model may inform the state,

tribe, or EPA in establishing NPDES permit requirements if the NPDES program is applicable. In many cases states will establish the authority for creation of management entities, provide funding, and provide technical assistance and training to local governments. The local governments would then have primary responsibility for implementation of the management program. If a decentralized system is required to have an NPDES permit and an authorized state or tribe is administering a decentralized management program under this strategy, the requirements of the program should be incorporated into the applicable NPDES permit, which is the primary regulatory instrument. If a state or tribe administering the program is not an authorized NPDES authority, the requirements of the program should be submitted to the NPDES permit issuing authority as a Section 401 certification requirement. If the program is being administered by a local authority or by a tribe without

401 certification ability, the requirements of the program should be recommended to the NPDES permit issuing authority for inclusion in the facilities permit. There are some cases, however, where the states themselves have the primary role and authority to implement the regulatory program at the local level.

Costs

State, tribal, and local governments must recognize that it is likely that both the regulatory authority and the property owner will face increased costs in improving management practices and programs. The cost impacts may increase as the level of management increases;

however, trade-offs exist. Costs incurred by the regulatory authority and/or management entity may be offset by increased permit fees and more efficient data management

tools, while the costs to the property owner may be offset by reduced repair and replacement costs, avoidance of environmental restoration costs, and increased property values and quality of life.

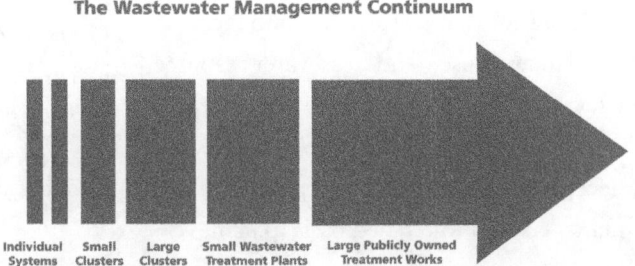

The Wastewater Management Continuum

Individual Systems | Small Clusters | Large Clusters | Small Wastewater Treatment Plants | Large Publicly Owned Treatment Works

Model 1 - The Homeowner Awareness Model

As a minimum level of management, EPA recommends **Model 1 - The Homeowner Awareness Model**. This program specifies appropriate management practices where treatment systems are owned and operated by individual property owners in areas of low environmental sensitivity, i.e., no restricting site or soil conditions such as shallow water tables or drinking water wells within locally determined horizontal setback distances. This model is applicable where treatment technologies are limited to conventional systems, which are passive and robust treatment systems that can provide acceptable treatment under suitable site conditions despite a lack of attention by the owner. Failures that might occur and continue undetected will pose a relatively low level of risk to public health and water resources. The objectives of this management model are to ensure that all systems are sited, designed, and constructed in compliance with sound, prevailing rules; all systems are documented and inventoried by the regulatory authority; and system owners are informed of the maintenance needs of their systems through timely reminders. The model is intended to provide an accurate record of the types and location of installed systems, to raise homeowners' awareness of basic system maintenance requirements, and to better ensure that the homeowners attend to those deficiencies that overtly threaten public health. This model, like all management programs described in this guidance, suggests the use of only trained and licensed/certified service providers. This model is a starting point for enhancing management programs because it provides communities with a good database of systems and their application for determining whether increased management practices are necessary.

Model 2 - The Maintenance Contract Model

EPA recommends **Model 2 - The Maintenance Contract Model** where more complex system designs are employed to enhance the capacity of conventional systems to accept and treat wastewater or where small clusters are used. For example, pretreating wastewater to remove nonbiodegradable materials and particulate matter that typically pass through a septic tank may enhance subsurface infiltration system performance on marginally suitable sites (sites with limited area, slowly permeable soils, or shallow water tables). However, such pretreatment units can have mechanical components and sensitive treatment processes, which require routine observation and maintenance if they are to perform satisfactorily. Maintenance of these more complex systems is critical to sustaining acceptable protection in these areas of greater environmental sensitivity. Therefore, these systems should be allowed only where trained operators are under contract to perform timely operation and maintenance. The objectives of this model build on the Homeowner Awareness Model by ensuring that property owners maintain maintenance contracts with trained operators.

Model 3 - The Operating Permit Model

EPA recommends **Model 3 - The Operating Permit Model** where sustained performance of onsite wastewater treatment systems is critical to protect public health and water quality. Examples of locations where this program might be appropriate include areas adjacent to estuaries or lakes where excessive nutrient concentrations may be a concern or situations where a source water assessment has identified onsite systems as potential threats to drinking water supplies. EPA strongly recommends that this be the minimum model used where large-capacity systems or systems treating high-strength wastewaters are present. EPA has determined not to regulate large-capacity onsite systems under the Underground Injection Control program at this time based on the belief that implementation of these Management Guidelines can ensure adequate protection of public health and the environment.[10] A principal objective of this management program is to ensure that the onsite wastewater treatment systems continuously meet their performance criteria. Limited-term operating permits are issued to the property owner and are renewable for another term if the owner demonstrates that the system is in compliance with the terms and conditions of the permit. In subareas where it is appropriate to use conventional onsite system designs, the operating permit may contain only a requirement that routine maintenance be performed in a timely manner and the condition of the system be inspected periodically. With complex systems, the treatment process will require more frequent inspections and adjustments, so process monitoring may be required. An advantage to implementing the program elements and activities of this management program is that the design of treatment systems is based on performance criteria that are less dependent on site characteristics and conditions. Therefore, systems can be used safely in more sensitive environments if their performance meets those requirements reliably and consistently. The operating permit provides a mechanism for continuous oversight of system performance and negotiating timely corrective actions or levying penalties if compliance with the permit is not maintained. To comply with these performance standards, the property owner should be encouraged to hire a licensed maintenance provider or operator.

Model 4 - The Responsible Management Entity (RME) Operation and Maintenance Model

EPA recommends **Model 4 - The Responsible Management Entity (RME) Operation and Maintenance Model** where large numbers of onsite and clustered systems must meet specific water quality requirements because the sensitivity of the environment is high, e.g., wellhead protection areas or shellfish waters. Frequent and highly reliable operation and maintenance is required to ensure water resource protection. Issuing the operating permit to an RME instead of the property owner provides greater assurance of control over performance compliance. This allows the use of performance-based systems in more sensitive environments than the Operating Permit Model. For a service fee, an RME takes responsibility for the operation and maintenance. This approach can reduce the number of permits and the administration functions performed by the regulatory authority. System failures are also reduced as a result of routine and preventive maintenance. The operating permit system is identical to that of the Operating Permit Model except that the permittee is a public or private RME. States may need to establish (and some already have) a regulatory structure to oversee the rate structures that RMEs establish and any other measures that a public services commission would normally undertake to manage private entities in noncompetitive situations.

Model 5 - The Responsible Management Entity (RME) Ownership Model

Model 5 - The Responsible Management Entity (RME) Ownership Model is a variation of the RME operation and maintenance concept in the RME Operation and Maintenance Model, with the exception that ownership of the system is no longer with the property owner. The designated management entity owns, operates, and manages the decentralized wastewater treatment systems in a manner analogous to central sewerage. Under this approach, the RME maintains control of planning and management, as well as operation and maintenance. This management model is appropriate for environmental or public health conditions similar to those for the RME Operation and Maintenance Model, but Model 5 provides a higher level of control of system performance. It also reduces the likelihood of disputes that can occur between the RME and the property owner in the RME Operation and Maintenance Model when the property owner fails to fully cooperate with the RME. The RME can also more readily replace existing systems with higher-performance units or clustered systems when necessary. EPA recommends implementation of the management practices detailed in the RME Ownership Model in cases such as where new, high-density development is proposed in the vicinity of sensitive receiving waters. States might need to establish a regulatory structure to oversee the rate structures that RMEs establish and any other measures that a public services commission would normally undertake to manage entities in noncompetitive situations.

HOW TO APPLY THE MANAGEMENT MODELS

Tables 1 through 5 in Appendix A provide brief descriptions of specific activities to be undertaken for the various program elements of a management model. The party that has primary responsibility for the activities is also identified. The program elements and activities listed for each management model are

considered to be the minimum elements and activities necessary to achieve the stated management objectives for each model. A detailed discussion of the program elements and activities is provided in the draft Management Handbook. The handbook complements the Management Guidelines and helps states, tribes, and local communities that wish to evaluate and upgrade their existing programs to develop and implement improved management programs. The draft Management Handbook includes case studies and examples of materials used by communities that have adequately implemented management programs.

How Do the Models Apply to Local Conditions?

As previously indicated, the management model a particular community or service area selects should be based on environmental sensitivity, public health risks, the complexities of the wastewater treatment technologies that might or should be implemented, and the size or density of development. The management model is selected after the decision to use decentralized wastewater treatment is made. The tables in Appendix A generally describe

Selection of the management model is made after the decision to use decentralized wastewater treatment is made.

recommended activities for each of the management elements associated with the management models. How each of these elements and activities will be implemented will depend on decisions by the local community and regulatory authority, based on generally accepted onsite wastewater science and practice, locally appropriate statutes, ordinances, institutional structures, technical capabilities, public preferences, and other factors. Thus, the general framework for a local management program should be derived from the tables, but it must be tailored to suit local circumstances and preferences.

The general framework for a local management program should be derived from the tables, but it must be tailored to suit local circumstances and preferences.

EPA recognizes the varied nature of management needed across the country and within states and localities, the need for flexibility in adopting the recommendations of the Management Guidelines, and the lack of resources for implementation. Although states, tribes, and local communities are encouraged to implement management models, an individual program may properly include elements of several management models. These hybrid or combination programs may be appropriate where site conditions vary within the community or institutional capacity is not uniform within the jurisdiction. It is also recommended that appropriate levels of management for decentralized systems be established in jurisdictions that have both centralized and decentralized wastewater treatment. In some cases, it might be feasible for the entity that manages the centralized wastewater treatment facility to manage the decentralized systems as well.

How Can a Community Phase In a Management Model?

Targeting of specific types of systems for improved management may also be appropriate when resources are limited and a phased approach that focuses on priority systems is preferred. When there are limited resources for monitoring efforts, a widely used approach has been to initially target higher-density or environmentally sensitive areas. Examples of environmentally sensitive areas include areas used for drinking water sources, areas adjacent to heavily used lakes and beaches, and areas that affect coral reefs or shellfish beds. Any approach taken should include input from all the stakeholders in a local jurisdiction or watershed.

The implementation of higher levels of management will often occur in progressive stages, as more performance data and experience with systems develop, public awareness and support increase, and the

capability of state, tribal, and local institutions to deal with management challenges builds over time. Implementation of the elements and activities recommended by the Homeowner Awareness Model as the threshold level of management will not only raise the quality of management practices for most existing programs but also initiate activities (such as an inventory of systems) that allow the community to identify and address circumstances that might require upgrading to higher levels of management.

Although the Homeowner Awareness Model might adequately address conventional systems within low-risk

segments of a service area, there might be other areas of higher risk that require higher levels of management. For those areas, a higher-level management model, more appropriate for areas with higher sensitivities, may be incorporated into the overall management program to customize system management to the needs of the community or service area. It is important that the management program be structured to adequately manage an appropriate set of onsite and clustered systems for the full range of environmental conditions. For example, the Operating Permit Model might be selected for the more sensitive areas such as those along lakefronts or estuaries shown to have poor water quality, while a lower-level management model might still be appropriate where the receiving environment is not as sensitive and conventional systems are acceptable.

> *Targeting of specific types of systems for improved management may also be appropriate when resources are limited.*

What Should Be Considered When Selecting a Management Model?

• *Environmental sensitivity and public health risk.* The locally developed management program should be based on the potential risk of onsite wastewater treatment system discharges adversely affecting the public health or the quality of local water resources. The level of oversight incorporated into the management program should increase as the potential for negative impacts on public health or for environmental degradation increases. Examples of parameters to consider in assessing public health and environmental sensitivity include soil permeability, depth to a restrictive horizon and ground water, aquifer type, receiving water use, proximity to surface waters, topography, geology, location of critical habitat under the Endangered Species Act, and density of development. Another useful parameter to consider is the "susceptibility determinations" that states and tribes will make as part of their source water assessments. These assessments determine which potential sources of pollution, including decentralized wastewater systems, pose the greatest threats to drinking water.

Other issues to consider that might have a direct impact on public health include the need to protect shellfish harvesting and direct contact recreational waters. An area with moderately permeable soils and a ground water table that is sufficiently isolated from

the effects of onsite discharges may be designated as an area of low public health risk and environmental sensitivity, whereas an area with excessively permeable soils with a shallow water table used for a drinking water source would be designated as an area of high concern. For those watersheds where a determination has been made that the onsite wastewater treatment system is contributing to a violation of water quality standards, the elements and activities of the Operating Permit Model, the RME Operation and Maintenance Model, or the RME Ownership Model should be selected to address restoration of the watershed. More detailed information on these factors is provided in the draft Management Handbook.

- *Complexity of treatment system.* The complexity of the treatment system also influences the management program selected. As the complexity of a treatment system increases to meet management objectives or system performance standards, the need for a higher level of operation and maintenance and monitoring increases to ensure that the system does not malfunction to create an unacceptable risk to public health or water resources. A less complex treatment system, such as a conventional onsite septic

system, depends upon passive, natural processes for the movement, treatment, and dispersal of wastewater. The prescriptive elements of the Homeowner Awareness Model, where properly applied, might be sufficient for conventional onsite technologies to consistently function as effective wastewater treatment systems. A more complex treatment system, such as a surface discharging aerobic treatment system with filtration and disinfection, will require routine monitoring and attention from a professional technician to maintain performance and therefore requires a higher level of oversight. EPA's updated *Onsite Wastewater Treatment Systems Manual*[11] provides guidance on performance and management requirements for a broad range of onsite treatment and dispersal technologies. System size also influences the management model selected. Large-capacity and clustered systems require a higher degree of management than individual onsite systems.

> *EPA's updated* Onsite Wastewater Treatment Systems Manual, *provides guidance on performance and management requirements for a broad range of onsite treatment and dispersal technologies.*

Communities that have made the decision to use onsite and clustered systems should use these Management Guidelines as a tool for identifying approaches for proper management of the systems. Implementation of the management practices defined in the Management Guidelines will help communities meet water quality and public health goals, provide a greater range of options for cost-effectively meeting wastewater needs, and protect consumers' investment in home and business ownership. Tables 1 through 5 in Appendix A provide a useful summary of the program elements for each management model and the associated responsible party and activity. The draft Management Handbook provides further detail on how to implement the management programs and is designed to assist state, tribal, and local officials; service providers; and other interested parties with improving system operation, maintenance, and performance.

Where Can Further Information Be Obtained?

Visit EPA's Web site on decentralized wastewater treatment at www.epa.gov/owm/onsite. The site includes a copy of the draft Management Handbook, fact sheets on technologies, useful links to other sites, a calendar of events, frequently asked questions, sources of funding information on demonstration projects, and numerous reference documents such as EPA's new *Onsite Wastewater Treatment Systems Manual*.

Additional copies of this document, (EPA 832-B-03-001), can be obtained from:

U.S. EPA Publications Clearinghouse
P.O. Box 42419
Cincinnati, OH 45242
Telephone: 800-490-9198
Fax: 513-489-8695

REFERENCES

1. U.S. Department of Commerce, U.S. Census Bureau, *American Housing Survey for the United States–1995*, issued September 1997.

2. V.I., Nelson, S.P. Dix, and F. Shephard, *Advanced Onsite Wastewater Treatment and Management Scoping Study: Assessment of Short-Term Opportunities and Long-Run Potential (DRAFT)*, May 1999. Data based on reporting from onsite system inspections in Massachusetts.

3. U.S. Environmental Protection Agency, *National Water Quality Inventory: 1996 Report to Congress*, EPA 841-R-97-008, 1998.

4. U.S. Environmental Protection Agency, *1996 Clean Water Needs Survey Report to Congress*, EPA/832/R-97/003, September 1997.

5. U.S. Department of Commerce, National Oceanic and Atmospheric Administration, *National Shellfish Register*, 1995.

6. Sarasota Bay (FL) National Estuary Program (www.sarasotabay.org), *Sarasota Bay: Reclaiming Paradise —A Vision for Sarasota Bay (State of the Bay, 1992)*.

7. U.S. Environmental Protection Agency, *40 CFR Parts 141 and 142: National Primary Drinking Water Regulations: Ground Water Rule; Proposed Rules*, Federal Register, May 10, 2000.

8. U.S. Centers for Disease Control, *Database of Waterborne and Foodborne Disease Outbreaks in the United States, 1971–1994*. Data summary and analysis provided by EPA as part of the development of the Ground Water Rule, and found at www.epa.gov/ogwdw000/standard/phs.html.

9. U.S. Environmental Protection Agency, *Response to Congress on Use of Decentralized Wastewater Treatment Systems*, EPA 832/R-97/001b, April 1997.

10. U.S. Environmental Protection Agency, *40 CFR Part 144, Underground Injection Control Program—Notice of Final Determination for Class V Wells; Final Rule*, Federal Register, June 7, 2002, Vol. 67, No. 110, pages 39584–39593.

11. U.S. Environmental Protection Agency, *Onsite Wastewater Treatment Systems Manual*, EPA 625/R-00/008, February 2002.

12. U.S. Environmental Protection Agency, *1998 Section 303(d) List Fact Sheet: National Picture of Impaired Waters Highlights of the 1998 303(d) Lists (based on Tracking System data available 04/06/00)*. Found at http://www.epa.gov/owow/tmdl/states/national.html

13. U.S. Environmental Protection Agency, *Domestic Septage Regulatory Guidance: A Guide to the EPA 503 Rule*, EPA 832/B-92/005, 1993.

14. U.S. Environmental Protection Agency, *Guide to Septage Treatment and Disposal*, EPA 625/R-94/002, 1994.

15. U.S. Environmental Protection Agency, *Draft Implementation Guidance for Ambient Water Quality Criteria for Bacteria–1986*, EPA 823/D-00/001, January 2000.

16. U.S. Environmental Protection Agency, *Nutrient Criteria Technical Guidance Manual—Estuarine and Coastal Marine Waters*, EPA 822/B-01/003, October 2001.

17. U.S. Environmental Protection Agency, *Nutrient Criteria Technical Guidance Manual—Lakes and Reservoirs*, EPA 822/B-00/001, April 2000.

18. U.S. Environmental Protection Agency, *Nutrient Criteria Technical Guidance Manual—Rivers and Streams*, EPA 822/B-00/002, July 2000.

19. U.S. Environmental Protection Agency, *Class V Underground Injection Control Study*, EPA 816/R-99/014, September 1999.

20. U.S. Environmental Protection Agency, *Guidance Specifying Management Measures for Sources of Nonpoint Pollution in Coastal Waters*.

21. U.S. Environmental Protection Agency, *Onsite Wastewater Treatment Systems Manual*, EPA 625/R-00/008, February 2002.

22. U.S. Environmental Protection Agency, *Design Manual— Onsite Wastewater Treatment and Disposal Systems*, EPA 625/1-80/012, October 1980.

GLOSSARY

Aerobic Treatment Unit (ATU): A mechanical wastewater treatment unit that provides secondary wastewater treatment for a single home, a cluster of homes, or a commercial establishment by mixing air (oxygen) and aerobic and facultative microbes with the wastewater. ATUs typically use a suspended growth process (such as activated sludge-extended aeration and batch reactors), a fixed-film process (similar to a trickling filter), or a combination of the two treatment processes.

Alternative Onsite Treatment System: A wastewater treatment system that includes components different from those typically used in a conventional septic tank and subsurface wastewater infiltration system (SWIS). An alternative system is used to achieve acceptable treatment and dispersal of wastewater where conventional systems either might not be capable of protecting public health and water quality or are inappropriate for properties with shallow soils over ground water or bedrock or soils with low permeability. Examples of components that can be used in alternative systems are sand filters, aerobic treatment units, disinfection devices, and alternative subsurface infiltration designs such as mounds, gravelless trenches, and pressure and drip distribution.

Centralized Wastewater System: A managed system consisting of collection sewers and a single treatment plant used to collect and treat wastewater from an entire service area. Traditionally, such a system has been called a publicly owned treatment works (POTW) as defined at 40 CFR 122.2.

Cesspool: A drywell that receives untreated sanitary waste containing human excreta, which sometimes has an open bottom and/or perforated sides (40 CFR 144.3).

Cesspools with the capacity to serve 20 or more persons per day were banned in federal regulations promulgated on December 7, 1999. The construction of new cesspools was immediately banned, and existing large-capacity cesspools must be replaced with sewer connections or onsite wastewater treatment systems by 2005.

Clustered System: A wastewater collection and treatment system under some form of common ownership that collects wastewater from two or more dwellings or buildings and conveys it to a treatment and dispersal system located on a suitable site near the dwellings or buildings.

Construction Permit: A permit issued by the designated local regulatory authority that allows the installation of a wastewater treatment system in accordance with approved plans and applicable codes.

Conventional Onsite Treatment System: A wastewater treatment system consisting of a septic tank and a typical trench or bed subsurface wastewater infiltration system.

Decentralized System: An onsite or clustered system used to collect, treat, and disperse or reclaim wastewater from a small community or service area.

Dispersal System: A system that receives pretreated wastewater and releases it into the air, into surface or ground water, or onto or under the land surface. A subsurface wastewater infiltration system is an example of a dispersal system.

Engineered Design: An onsite or clustered wastewater system that is designed and certified by a licensed/certified designer to meet specific performance criteria for a particular wastewater on a particular site.

Environmental Sensitivity: The relative susceptibility to adverse impacts of a water resource or other receiving environment from dispersal of wastewater or its constituents. The impacts may be low, acute (immediate and significantly disruptive), or chronic (long-term, with gradual but serious disruptions).

Large-Capacity Septic System: An onsite method of partially treating and disposing of sanitary wastewater having the capacity to serve 20 or more persons per day subject to EPA's Underground Injection Control regulations.

Management Model: A 13-element program designed to protect and sustain public health and water quality through the use of appropriate policies and administrative procedures that define and integrate the roles and responsibilities of the regulatory authority, system owner, service providers, and management entity, when present, to ensure that onsite and clustered wastewater treatment systems are appropriately managed throughout their life cycle. The program elements include public education and participation; planning; performance; training and certification/licensing; site evaluation; design; construction; operation and maintenance; residuals management; compliance inspections/monitoring; corrective actions; recordkeeping, inventory, and reporting; and financial assistance and funding. Management services should be provided by properly trained and certified personnel and tracked through a comprehensive management information system.

National Pollutant Discharge Elimination System (NPDES): A national program under Section 402 of the Clean Water Act for regulation of discharges of pollutants from point sources to waters of the United States. Discharges are illegal unless authorized by an NPDES permit.

Onsite Service Provider: A person who provides onsite system services. Providers include (but are not limited to) designers, engineers, soil scientists, site evaluators, installers, contractors, operators, managers, maintenance-service providers, pumpers, and others who provide services to system owners or other service providers.

Onsite Wastewater Treatment System (OWTS): A system relying on natural processes and/or mechanical components to collect, treat, and disperse or reclaim wastewater from a single dwelling or building.

Operating Permit: A renewable and revocable permit to operate and maintain an onsite or clustered treatment system in compliance with specific operational or performance criteria stipulated by the regulatory authority.

Performance-Based Management Program: A program designed to preserve and protect public health and water quality by seeking to ensure sustained achievement of specific, measurable performance criteria based on site and risk assessments.

Performance Criteria: Any criteria established by the regulatory authority to ensure future compliance with the public health and water quality goals of the community, the state or tribe, and the federal government. Performance criteria can be expressed as numeric limits (e.g., pollutant concentrations, mass loads, wet weather flow, structural strength) or narrative descriptions of desired conditions or requirements (e.g., no visible scum, sludge, sheen, odors, cracks, or leaks).

Permitting Authority: The state, tribal, or local unit of government with the statutory or delegated authority to issue permits to build and operate onsite wastewater systems.

Prescription-Based Management Program: A program designed to preserve and protect public health and water quality by specifying preengineered system designs for specific sets of site conditions such that systems that are sited, designed, and constructed properly are deemed to meet public health and water quality standards.

Prescriptive Requirements: Specifications for design, installation, and other procedures and practices for onsite or clustered wastewater systems on sites that meet stipulated criteria. Proposed deviations from the stipulated criteria, specifications, procedures, or practices require formal approval from the regulatory authority.

Regulatory Authority (RA): The unit of government that establishes and enforces codes related to the permitting, design, placement, installation, operation, maintenance, monitoring, and performance of onsite and clustered wastewater systems.

Residuals: The solids generated or retained during the treatment of wastewater. They include trash, rags, grit, sediment, sludge, biosolids, septage, scum, and grease, as well as those portions of treatment systems that have served their useful life and require disposal, such as the sand or peat from a filter. Because of the different characteristics of residuals, management requirements can differ as stipulated by the appropriate federal regulations.

Responsible Management Entity (RME): A legal entity responsible for providing various management services with the requisite managerial, financial, and technical capacity to ensure the long-term, cost-effective management of decentralized onsite or clustered wastewater treatment facilities in accordance with applicable regulations and performance criteria.

Septage: The liquid and solid materials pumped from a septic tank during cleaning operations.

Septic Tank: A buried, watertight tank designed and constructed to receive and partially treat raw wastewater. The tank separates and retains settleable and floatable solids suspended in the wastewater and discharges the settled wastewater for further treatment and dispersal to the environment.

Source Water Assessment: A study and report required by the Source Water Assessment Program (SWAP) of the Safe Drinking Water Act addressing the capability of a given public water system to protect water quality. The assessment includes delineation of the source water area, identification of potential sources of contamination in the delineated area, determination of susceptibility to those sources, and public notice of the completed assessment.

Underground Injection Well: A constructed system designed to place waste fluids above, into, or below aquifers classified as underground sources of drinking water. As regulated under the Underground Injection Control (UIC) Program of the Safe Drinking Water Act (40 CFR Parts 144 and 146), injection wells are grouped into five classes. Class V includes shallow systems such as cesspools and subsurface wastewater infiltration systems. Subsurface wastewater infiltration systems with the capacity to serve 20 or more people per day, or similar systems receiving nonsanitary wastes, are subject to federal regulation. Class V motor vehicle waste injection wells and large-capacity cesspools are specifically prohibited under the UIC regulations.

APPENDIX A: MANAGEMENT MODELS

This appendix presents a description of activities associated with each program element and identifies the party responsible for each activity. A detailed discussion is presented in the Management Handbook. Activities in bold are activities added to program elements from the preceding Management Model.

Note: If applicable, National Pollutant Discharge Elimination System (NPDES) requirements under the Clean Water Act (CWA) or Underground Injection Control (UIC) requirements under the Safe Drinking Water Act (SDWA) supercede any less stringent or inconsistent provisions. Program elements in each model help inform the state, tribe, or EPA in establishing NPDES permit requirements.

MANAGEMENT MODEL 1: HOMEOWNER AWARENESS

Objective: To ensure that conventional onsite systems are sited and constructed properly in accordance with appropriate state, tribal, and local regulations and codes; that they are periodically inspected; and, if necessary, that they are repaired by the Owner. The Regulatory Authority maintains a record of the location of all systems and periodically provides the Owner/User with notices regarding operation and preventive maintenance recommendations.

PROGRAM ELEMENT	RESPONSIBLE PARTY	ACTIVITY
PUBLIC EDUCATION AND PARTICIPATION	Regulatory Authority	• Educate Owner/User on purpose, use, and care of treatment system. • Provide public review and comment periods of any proposed program or rule changes.
	Service Provider	• Be informed of existing rules and review and comment on any proposed program and/or rule changes. • Participate in advisory committees established by the Regulatory Authority.
	Owner/User	• Be informed of purpose, use, and care of treatment system. • Be informed of existing rules and review and comment on any proposed program and/or rule changes. • Participate in advisory committees established by the Regulatory Authority.
PLANNING	Regulatory Authority	• Coordinate program rules and regulations with state, tribal, and local planning and zoning and other water-related programs. • Evaluate potential risks of wastewater discharges to limit environmental impacts on receiving environments during the rule making process. • Limit potential risks of environmental impacts from residuals management program and evaluate available handling/treatment capacities. • Inform local planning authority of rule changes and recommend its evaluation of potential impacts on land use.
	Developer	• Hire planners, certified site evaluators, and designers to ensure that all lots of proposed subdivision plats meet requirements for onsite treatment prior to final plat.
PERFORMANCE	Regulatory Authority	• Establish system failure criteria to protect public health, e.g., wastewater backups in building, wastewater ponding on ground surface, insufficient separation from ground water or wells.
	Owner/User	• Regularly maintain system in proper working order.
TRAINING AND CERTIFICATION/ LICENSING	Licensing Board/ Regulatory Authority	• Develop and administer training, testing, and certification/licensing program for site evaluators, designers, contractors, and pumpers/haulers. • Maintain a current certified/licensed Service Provider listing.
	Service Provider	• Obtain appropriate certification(s)/license(s) and continuing education as required. • Obtain training from the manufacturer or vendor regarding appropriate use, installation requirements, and O&M procedures of any proprietary equipment to be installed. • Comply with applicable federal, state, tribal, and local requirements.
	Owner/User	• When using third-party services, contract with only the appropriate certified/licensed Service Providers.
SITE EVALUATION	Regulatory Authority	• Codify prescriptive requirements for site evaluation procedures. • Codify criteria for treatment site characteristics suitable for permitted designs that will prevent unacceptable impacts on ground and surface water resources.
	Site Evaluator	• Obtain certification/license to practice. • Describe site and soil characteristics, determine suitability of site with respect to code requirements, and estimate site's hydraulic and treatment capacity. • Comply with applicable federal, state, tribal, and local requirements in the evaluation of sites for wastewater treatment and dispersal.
	Owner	• Hire a certified/licensed site evaluator to perform site evaluation.

MANAGEMENT MODEL 1: HOMEOWNER AWARENESS

MANAGEMENT MODEL 1: HOMEOWNER AWARENESS

PROGRAM ELEMENT	RESPONSIBLE PARTY	ACTIVITY
DESIGN	Regulatory Authority	• Codify prescriptive, preengineered designs that are suitable for treatment sites that meet the appropriate prescriptive site criteria.
	Designer	• Obtain a certification/license to practice. • Design a treatment system that is compatible with the site and soil characteristics described by the site evaluator. • Comply with applicable federal, state, tribal, and local requirements in the design of wastewater treatment and dispersal systems.
	Owner	• Hire a certified/licensed designer to prepare system design.
CONSTRUCTION	Regulatory Authority	• Administer a permitting program for system construction, including Regulatory Authority review of proposed system siting and design plans. • Perform final construction inspection for compliance assurance and inventory data collection. • Require that record drawings of constructed system be submitted to the Regulatory Authority by Owner.
	Contractor/ Installer	• Obtain certification/license to practice. • Construct the system in accordance with the approved plans and specifications. • Prepare record drawings of completed system and submit to Owner. • Comply with applicable federal, state, tribal, and local requirements in the design and construction of wastewater treatment and dispersal systems.
	Designer of Record	• Approve proposed field changes and submit to Owner. • Comply with applicable federal, state, tribal, and local requirements in the design and construction of wastewater treatment and dispersal systems.
	Owner	• Hire a certified/licensed contractor/installer to construct system. • Submit final record drawings of constructed system to Regulatory Authority.
OPERATION & MAINTENANCE	Regulatory Authority	• Provide Owner/User with educational materials regarding system use and care. • Send timely reminder to Owner of when scheduled preventive maintenance is due.
	Pumper/Hauler	• Obtain certification/license to practice. • Inspect and service system as necessary. • Comply with applicable federal, state, tribal, and local requirements in the operation and maintenance of the treatment and dispersal system.
	Owner	• Perform recommended routine maintenance or hire a certified/licensed pumper/hauler to perform maintenance. • Hire a certified/licensed pumper/hauler to periodically inspect, service, and remove septage for proper treatment and disposal.
	User	• Follow recommendations provided by Regulatory Authority, Service Providers, and/or Owner to ensure that undesirable or prohibited materials are not discharged to system.
RESIDUALS MANAGEMENT	Regulatory Authority	• Administer a tracking system for residuals hauling, treatment, and disposal and review to evaluate compliance with 40 CFR Part 503 (Use and Disposal of Sewage Sludge), 40 CFR Part 257, and applicable state, tribal, and local requirements. • Inventory available residuals handling/treatment capacities and develop contingency plans to ensure that sufficient capacities are always available.
	Pumper/Hauler	• Obtain certification/license to practice. • Comply with applicable federal, state, tribal, and local requirements in the pumping, hauling, treatment, and disposal of treatment system residuals.
COMPLIANCE INSPECTIONS/ MONITORING	Regulatory Authority	• Conduct final construction inspections to ensure compliance with approved plans and permit requirements. • Perform compliance inspections at point-of-sale, change-in-use of properties, "targeted areas," and systems reported to be in violation. • Conduct compliance inspections of residuals hauling, treatment, and disposal.
	Pumper/Hauler	• Inform Owner of any noncompliant items observed during routine servicing of system.
	Owner	• Periodically perform a "walk-over" inspection of the system and correct any deficiencies.

PROGRAM ELEMENT	RESPONSIBLE PARTY	ACTIVITY
CORRECTIVE ACTIONS	Regulatory Authority	• Negotiate compliance schedule with Owner for correcting documented noncompliance items. • Administer enforcement program, including fines and/or penalties for failure to comply with compliance requirements. • Obtain necessary authority to enter property to correct imminent threats to public health if the Owner/User fails to comply.
	Designer	• Provide Owner with documents (drawings, specifications, modifications, etc.) that may be required by Regulatory Authority prior to corrective action.
	Contractor/ Installer	• Perform required repairs, modifications, and upgrades as necessary.
	Owner	• Comply with terms and conditions of the negotiated compliance schedule. • Submit required documents for corrective actions to Regulatory Authority. • Hire appropriate certified/licensed Service Providers to perform required corrective actions.
RECORD KEEPING, INVENTORY, & REPORTING	Regulatory Authority	• Administer a database inventory (locations, site evaluations, record drawings, permits, performed maintenance, inspection reports) of all systems. • Maintain a residuals treatment and disposal tracking system. • Maintain a current certified/licensed Service Provider listing that is available to the public.
	Pumper/Hauler	• Prepare and submit records of residuals handling as required.
	Owner	• Maintain approved record drawings of system. • Maintain maintenance records of system. • Provide drawings, specifications, and maintenance records to new property owner at time of property transfer.
FINANCIAL ASSISTANCE & FUNDING	Regulatory Authority	• Provide the legal and financial support to sustain the management program. • Provide a listing of financial assistance programs available to Owner and the qualifying criteria for each program. • Consider implementing a state or local financing program to assist Owners in upgrading their systems.

MANAGEMENT MODEL 1: HOMEOWNER AWARENESS

MANAGEMENT MODEL 2: MAINTENANCE CONTRACTS

Objective: To allow use of more complex mechanical treatment options or small clusters through the requirement that maintenance contracts be maintained between the Owner and maintenance provider to ensure appropriate and timely system component maintenance by qualified technicians over the service life of the system.

PROGRAM ELEMENT	RESPONSIBLE PARTY	ACTIVITY[1]
PUBLIC EDUCATION AND PARTICIPATION	Regulatory Authority	• Educate Owner/User on purpose, use, and care of treatment system. • Provide public review and comment periods of any proposed program and/or rule changes.
	Service Provider	• Be informed of existing rules, and review and comment on any proposed program or rule changes. • Participate in advisory committees established by the Regulatory Authority.
	Owner/User	• Be informed of purpose, use, and care of treatment system. • Be informed of existing rules, and review and comment on any proposed program or rule changes. • Participate in advisory committees established by the Regulatory Authority.
PLANNING	Regulatory Authority	• Coordinate program rules and regulations with state, tribal, local planning and zoning and other water-related programs. • Evaluate potential risks of wastewater discharges to limit environmental impacts on receiving environments during the rule making process. • Limit potential risks of environmental impacts from residuals management program and evaluate available handling/treatment capacities. • Inform local planning authority of rule changes and recommend its evaluation of potential impacts on land use.
	Developer	• Hire planners, certified site evaluators, and designers to ensure that all lots of proposed subdivision plats meet requirements for onsite treatment prior to final plat.
PERFORMANCE	Regulatory Authority	• Establish system failure criteria to protect public health, e.g., wastewater backups in building, wastewater ponding on ground surface, insufficient separation from ground water or wells. • **Establish minimum performance criteria for manufactured component approvals.** • **Establish minimum maintenance requirements for approved systems.**
	Owner/User	• Regularly maintain system in proper working order.
TRAINING AND CERTIFICATION/ LICENSING	Licensing Board/ Regulatory Authority	• Develop and administer training, testing, and certification/licensing program for site evaluators, designers, contractors, operators, and pumpers/haulers. • Maintain a current certified/licensed Service Provider listing.
	Service Provider	• Obtain appropriate certification(s)/license(s) and continuing education as required. • Obtain training from the manufacturer or vendor regarding appropriate use, installation requirements, and O&M procedures of any proprietary equipment to be installed. • Comply with applicable federal, state, tribal, and local requirements.
	Owner/User	• When using third-party services, contract only with the appropriate certified/licensed Service Providers.
SITE EVALUATION	Regulatory Authority	• Codify prescriptive requirements for site evaluation procedures. • Codify criteria for treatment site characteristics suitable for permitted designs that will prevent unacceptable impacts on ground and surface water resources. • **Establish alternative site acceptance criteria for approved systems providing enhanced pretreatment.**
	Site Evaluator	• Obtain certification/license to practice. • Describe site and soil characteristics, determine suitability of site with respect to code requirements, and estimate site's hydraulic and treatment capacity. • Comply with applicable federal, state, tribal, and local requirements in the evaluation of sites for wastewater treatment and dispersal.
	Owner	• Hire a certified/licensed site evaluator to perform site evaluation.

[1] *Activities in bold are activities added to program elements from the preceding Management Model.*

MANAGEMENT MODEL 2: MAINTENANCE CONTRACTS

PROGRAM ELEMENT	RESPONSIBLE PARTY	ACTIVITY[1]
DESIGN	Regulatory Authority	• Codify prescriptive, preengineered designs that are suitable for treatment sites that meet the appropriate prescriptive site criteria. • **Administer an evaluation program for approving manufactured components for use with pre-engineered designs.**
	Designer	• Obtain certification/license to practice. • Design a treatment system that is compatible with the site and soil characteristics described by the site evaluator. • Comply with applicable federal, state, tribal, and local requirements in the design of wastewater treatment and dispersal systems.
	Owner	• Hire a certified/licensed designer to prepare system design.
CONSTRUCTION	Regulatory Authority	• Administer a permitting program for system construction, including Regulatory Authority review of proposed system siting and design plans. • Perform final construction inspection for compliance assurance and inventory data collection. • Require that record drawings of constructed system be submitted to the Regulatory Authority by Owner. • **Require Owner to submit a copy of system O&M manual to the Regulatory Authority.**
	Contractor/ Installer	• Obtain certification/license to practice. • Construct the system in accordance with the approved plans and specifications. • Prepare record drawings of completed system and submit to Owner. • **Provide Owner with an O&M manual describing component manufacturer's maintenance and troubleshooting requirements/recommendations.** • Comply with applicable federal, state, tribal, and local requirements in the design and construction of wastewater treatment and dispersal systems.
	Designer of Record	• Approve proposed field changes and submit to Owner. • Comply with applicable federal, state, tribal, and local requirements in the design and construction of wastewater treatment and dispersal systems.
	Owner	• Hire a certified/licensed contractor/installer to construct system. • Submit final record drawings of constructed system to Regulatory Authority. • **Submit a copy of system O&M manual to Regulatory Authority to record required maintenance.**
OPERATION & MAINTENANCE	Regulatory Authority	• Provide Owner/User with educational materials regarding system use and care. • Send timely reminder to Owner when scheduled preventive maintenance is due. • **Administer a program that requires the Owner to attest periodically that he or she holds a valid contract with a certified/licensed operator to perform scheduled and any necessary maintenance according to the maintenance requirements described in submitted O&M manual.** • **Require Owner to submit a maintenance report signed/sealed by certified/licensed operator immediately following scheduled maintenance.**
	Operator	• Obtain certification/license to practice. • **Inspect and service system as necessary in accordance with the submitted O&M manual.** • **Certify to Owner that the required maintenance was performed in a timely manner, describing any system deficiencies observed.** • **Comply with applicable federal, state, tribal, and local requirements in the operation and maintenance of the treatment and dispersal system.**
	Pumper/Hauler	• Obtain certification/license to practice. • Inspect and service system as necessary. • Comply with applicable federal, state, tribal, and local requirements in the operation and maintenance of treatment and dispersal system.
	Owner	• Hire a certified/licensed pumper/hauler to periodically inspect, service, and remove septage or other residuals for proper treatment and disposal. • **Maintain contractual agreement with a certified/licensed operator to perform scheduled maintenance as required.** • **Inform Regulatory Authority of any change in maintenance contract status.**
	User	• Follow recommendations provided by Regulatory Authority, Service Providers, and/or Owner to ensure that undesirable or prohibited materials are not discharged to system.

[1] *Activities in bold are activities added to program elements from the preceding Management Model.*

MANAGEMENT MODEL 2: MAINTENANCE CONTRACTS

MANAGEMENT MODEL 2: MAINTENANCE CONTRACTS

PROGRAM ELEMENT	RESPONSIBLE PARTY	ACTIVITY[1]
RESIDUALS MANAGEMENT	Regulatory Authority	• Administer a tracking system for residuals hauling, treatment, and disposal and review to evaluate compliance with 40 CFR Part 503 (Use and Disposal of Sewage Sludge), 40 CFR Part 257, and applicable state, tribal, and local requirements. • Inventory available residuals handling/treatment capacities and develop contingency plans to ensure that sufficient capacities are always available.
	Pumper/Hauler	• Comply with applicable federal, state, tribal, and local requirements in the pumping, hauling, treatment, and disposal of treatment system residuals.
COMPLIANCE INSPECTIONS/ MONITORING	Regulatory Authority	• Conduct final construction inspections to ensure compliance with approved plans and permit requirements. • Perform compliance inspections at point-of-sale, change-in-use of properties, "targeted areas," and/or systems reported to be in violation. • Conduct compliance inspections of residuals hauling, treatment, and disposal. • **Administer program for confirming that Owners hold valid maintenance contracts with certified/licensed operators and for monitoring timely submittals of certified maintenance reports.**
	Operator or Pumper/Hauler	• Inform Owner of any noncompliant items observed during routine servicing of system.
	Owner	• Periodically perform a "walk-over" inspection of the system and correct any deficiencies. • **Attest to the Regulatory Authority that a valid contract exists with a certified/licensed operator to perform necessary system maintenance.** • **Submit a maintenance report signed/sealed by a certified/licensed Service Provider immediately following scheduled maintenance.**
CORRECTIVE ACTIONS	Regulatory Authority	• Negotiate compliance schedule with Owner for correcting documented noncompliant items. • Administer enforcement program, including fines and/or penalties for failure to comply with compliance requirements. • Obtain necessary authority to enter property to correct imminent threats to public health if the Owner/User fails to comply.
	Designer	• Provide Owner with documents (drawings, specifications, modifications, etc.) that may be required by Regulatory Authority prior to corrective action.
	Contractor/ Installer	• Perform required repairs, modifications, and upgrades as necessary.
	Owner	• Comply with terms and conditions of the negotiated compliance schedule. • Submit required documents for corrective actions to Regulatory Authority. • Hire appropriate certified/licensed Service Providers to perform required corrective actions.
RECORD KEEPING, INVENTORY, & REPORTING	Regulatory Authority	• Administer a database inventory (locations, site evaluations, record drawings, permits, performed maintenance, inspection reports) of all systems. • Maintain a residuals treatment and disposal tracking system. • Maintain a current certified/licensed Service Provider listing that is available to the public. • **Administer an Owner/Service Provider maintenance contract compliance and certified maintenance report tracking system.** • **Record maintenance contract requirement on property deed.** • **Administer a certified maintenance report tracking system.**
	Operator	• Provide certified report of all maintenance and observed system deficiencies to Owner.
	Pumper/Hauler	• Prepare and submit records of residuals handling as required.
	Owner	• Maintain approved record drawings and **O&M manual** of system. • Maintain maintenance records of system. • Provide drawings, specifications, **O&M manual**, and maintenance records to new property owner at time of property transfer.
FINANCIAL ASSISTANCE & FUNDING	Regulatory Authority	• Provide the legal and financial support to sustain the management program. • Provide a listing of financial assistance programs available to Owner/User and the qualifying criteria for each program. • Consider implementing a state or local financing program to assist Owners in upgrading their systems.

[1] *Activities in bold are activities added to program elements from the preceding Management Model.*

MANAGEMENT MODEL 3: OPERATING PERMITS

Objective: To issue renewable/revocable operating permits to system Owner that stipulate specific and measurable performance criteria for the treatment system and periodic submittals of compliance monitoring reports. The performance criteria are based on risks to public health and water resources posed by wastewater dispersal in the receiving environment. Operating permits allow the use of clustered or onsite systems on sites with a greater range of site characteristics.

PROGRAM ELEMENT	RESPONSIBLE PARTY	ACTIVITY[1]
PUBLIC EDUCATION AND PARTICIPATION	Regulatory Authority	• Educate Owner/User on purpose, use, and care of treatment system. • Provide public review and comment periods of any proposed program and/or rule changes.
	Service Provider	• Be informed of existing rules, and review and comment on any proposed program or rule changes. • Participate in advisory committees established by the Regulatory Authority.
	Owner/User	• Be informed of purpose, use, and care of treatment system. • Be informed of existing rules, and review and comment on any proposed program or rule changes. • Participate in advisory committees established by the Regulatory Authority.
PLANNING	Regulatory Authority	• Coordinate program rules and regulations with state, tribal, and local planning and zoning and other water-related programs. • Evaluate potential risks of wastewater discharges to limit environmental impacts on receiving environments during the rule making process. • Limit potential risks of environmental impacts from residuals management program and evaluate available handling/treatment capacities. • Inform local planning authority of rule changes and recommend its evaluation of potential impacts on land use.
	Developer	• Hire planners, certified site evaluators, and designers to ensure that all lots of proposed subdivision plats meet requirements for onsite treatment prior to final plat.
PERFORMANCE	Regulatory Authority	• Establish system failure criteria to protect public health, e.g., wastewater backups in building, wastewater ponding on ground surface, insufficient separation from ground water or wells. • Establish minimum maintenance requirements for approved systems. • **Establish performance criteria necessary to protect public health and water resources for each defined receiving environment in Regulatory Authority's jurisdiction.**
	Owner/User	• Operate and regularly maintain system in proper working order. • **Operate system to comply with performance criteria stipulated in operating permit.**
TRAINING AND CERTIFICATION/ LICENSING	Licensing Board/ Regulatory Authority	• Develop and administer a training, testing, and certification/licensing program for site evaluators, designers, contractors, operators, pumpers/haulers, and **inspectors**. • Maintain a current certified/licensed Service Provider listing.
	Service Provider	• Obtain appropriate certification(s)/license(s) and continuing education as required. • Obtain training from the manufacturer or vendor regarding appropriate use, installation requirements, and O&M procedures of any proprietary equipment to be installed. • Comply with applicable federal, state, tribal, and local requirements.
	Owner/User	• When using third-party services, contract with only the appropriate certified/licensed Service Providers.
SITE EVALUATION	Regulatory Authority	• Codify prescriptive requirements for site evaluation procedures. • Codify criteria for treatment site characteristics suitable for permitted designs that will prevent unacceptable impacts on ground and surface water resources. • **Establish defining characteristics for each receiving environment in the Regulatory Authority's jurisdiction.**
	Site Evaluator	• Obtain certification/license to practice. • Describe site and soil characteristics, determine suitability of site with respect to code requirements, and estimate site's hydraulic and treatment capacity. • Comply with applicable federal, state, tribal, and local requirements in the evaluation of sites for wastewater treatment and dispersal.
	Owner	• Hire a certified/licensed site evaluator to perform site evaluation.

[1] *Activities in bold are activities added to program elements from the preceding Management Model.*

MANAGEMENT MODEL 3: OPERATING PERMITS

MANAGEMENT MODEL 3: OPERATING PERMITS

PROGRAM ELEMENT	RESPONSIBLE PARTY	ACTIVITY[1]
DESIGN	Regulatory Authority	• Codify prescriptive, preengineered designs that are suitable for treatment sites that meet the appropriate prescriptive site criteria. • **Administer a plan review program for engineered designs to meet stipulated performance criteria.** • **Require submission of routine operation and emergency contingency plans that will sustain system performance and avoid unpermitted discharges.**
	Designer	• Obtain certification/license to practice. • Certified/licensed designer to design treatment system that is compatible with the site and soil characteristics described by the site evaluator. • Comply with applicable federal, state, tribal, and local requirements in the design of wastewater treatment and dispersal systems.
	Owner	• Hire a certified/licensed designer to prepare system design.
CONSTRUCTION	Regulatory Authority	• Administer a permitting program for system construction, including Regulatory Authority review of proposed system siting and design plans. • **Require designer of record to certify that completed system construction is in substantial compliance with approved plans and specifications.** • Require that record drawings of constructed system be submitted to the Regulatory Authority by Owner. • Require Owner to submit a copy of system O&M manual to the Regulatory Authority.
	Contractor/Installer	• Obtain certification/license to practice. • Construct the system in accordance with the approved plans and specifications. • Prepare record drawings of completed system and submit to Owner. • Provide Owner with an O&M manual describing component manufacturer's maintenance and troubleshooting requirements/recommendations. • Comply with applicable federal, state, tribal, and local requirements in the design and construction of wastewater treatment and dispersal systems.
	Designer of Record	• Approve proposed field changes and submit to Owner. • **Certify that construction of the system is substantially in conformance with the approved plans and specifications.**
	Owner	• Hire a certified/licensed contractor/installer to construct system. • Submit final record drawings of constructed system to Regulatory Authority. • Submit a copy of system O&M manual to Regulatory Authority to record required maintenance.
OPERATION & MAINTENANCE	Regulatory Authority	• Provide Owner/User with educational materials regarding system use and care. • **Administer a program of renewable/revocable operating permits that are issued to Owner stipulating system performance criteria, compliance monitoring reporting schedule, term of permit, and renewal option upon documented compliance with permit.** • **Track and review compliance monitoring reports to ensure that systems are operating in accordance with operating permits.**
	Operator	• Obtain certification/license to practice. • **Inspect and service system as necessary in accordance with the submitted O&M manual and/or operating permit stipulations.** • Certify to Owner that the required maintenance was performed in a timely manner, describing any system deficiencies observed. • Comply with applicable federal, state, tribal, and local requirements in the operation and maintenance of the treatment and dispersal system.
	Pumper/Hauler	• Obtain certification/license to practice. • Inspect and service system as necessary. • Comply with applicable federal, state, tribal, and local requirements in the operation and maintenance of the treatment and dispersal system.
	Owner	• Hire a certified/licensed pumper/hauler or operator to maintain system. • Maintain system in proper working order. • **Operate and maintain the system in accordance with O&M manual and/or operating permit stipulations.** • **Submit compliance monitoring reports to the Regulatory Authority according to the schedule stipulated in the operating permit.**
	User	• Follow recommendations provided by Regulatory Authority and/or Service Providers to ensure that undesirable or prohibited materials are not discharged to system.

[1] *Activities in bold are activities added to program elements from the preceding Management Model.*

PROGRAM ELEMENT	RESPONSIBLE PARTY	ACTIVITY[1]
RESIDUALS MANAGEMENT	Regulatory Authority	• Administer a tracking system for residuals hauling, treatment, and disposal and review to evaluate compliance with 40 CFR Part 503 Use and Disposal of Sewage Sludge, 40 CFR Part 257, and applicable state, tribal, and local requirements. • Inventory available residuals handling/treatment capacities and develop contingency plans to ensure that sufficient capacities are always available.
	Pumper/Hauler	• Comply with applicable federal, state, tribal, and local requirements in the pumping, hauling, treatment, and disposal of treatment system residuals.
COMPLIANCE INSPECTIONS/ MONITORING	Regulatory Authority	• Perform inspection programs at point-of-sale, change-in-use of properties, "targeted areas," and/or systems reported to be in violation. • Conduct compliance inspections of residuals hauling, treatment, and disposal. • **Administer a program to monitor timely submittals of acceptable compliance maintenance reports.** • **Notify Owner of impending scheduled submittals of compliance monitoring reports.** • **Perform system inspections randomly and/or at time of operating permit renewal.**
	Operator or Pumper/Hauler	• Inform Owner of any noncompliant items observed during routine servicing of system.
	Owner	• Submit compliance monitoring reports to Regulatory Authority as stipulated in operating permit. • **Submit compliance inspection report signed/sealed by a certified/licensed inspector prior to applying for renewal of operating permit.**
CORRECTIVE ACTIONS	Regulatory Authority	• Negotiate compliance schedule with Owner for correcting documented noncompliant items. • Administer enforcement program including fines and/or penalties for failure to comply with compliance requirements. • Obtain necessary authority to enter property to correct imminent threats to public health if the Owner/User fails to comply. • **Require system inspection by certified inspector at time of operating permit renewal.**
	Designer	• Provide Owner with documents (drawings, specifications, modifications, etc.) that may be required by Regulatory Authority prior to corrective action.
	Contractor/ Installer	• Perform required repairs, modifications, and upgrades as necessary.
	Inspector	• **Obtain certification/license to practice.** • **Inspect treatment system for compliance with operating permit prior to permit renewal.**
	Owner	• Comply with terms and conditions of the negotiated compliance schedule. • Submit required documents for corrective actions to Regulatory Authority. • Hire appropriate certified/licensed Service Providers to perform required corrective actions.
RECORD KEEPING, INVENTORY, & REPORTING	Regulatory Authority	• Administer a database inventory (locations, site evaluations, record drawings, permits, performed maintenance, and inspection reports) of all systems. • Maintain a residuals treatment and disposal tracking system. • Maintain a current certified/licensed Service Provider listing that is available to the public. • **Administer a tracking system for operating permits.** • **Administer a tracking database for compliance reports.**
	Operator or Inspector	• Provide certified report of all maintenance and observed system deficiencies to Owner. • **Perform system monitoring as stipulated in Owner's operating permit.**
	Pumper/Hauler	• Prepare and submit records of residuals handling as required.
	Owner	• Maintain approved record drawings and O&M manual of system. • Maintain maintenance records of system. • Submit compliance monitoring reports to Regulatory Authority. • Provide drawings, specifications, O&M manual, and maintenance records to new property owner at time of property transfer.
FINANCIAL ASSISTANCE & FUNDING	Regulatory Authority	• Provide the legal and financial support to sustain the management program. • Provide a listing of financial assistance programs available to Owner/User and the qualifying criteria for each program. • Consider implementing a state or local financing program to assist Owners in upgrading their systems.

MANAGEMENT MODEL 3: OPERATING PERMITS

[1] *Activities in bold are activities added to program elements from the preceding Management Model.*

MANAGEMENT MODEL 4: RME OPERATION AND MAINTENANCE

Objective: To ensure that onsite/decentralized systems consistently meet their stipulated performance criteria through Responsible Management Entities that are responsible for operation and performance of systems within their service areas.

PROGRAM ELEMENT	RESPONSIBLE PARTY	ACTIVITY[1]
PUBLIC EDUCATION AND PARTICIPATION	Regulatory Authority	• Educate Owner/User on purpose, use, and care of treatment system. • Hold public meetings to inform the public of any proposed program and/or rule changes.
	Service Provider	• Be informed of existing rules, and review and comment on any proposed program or rule changes. • Participate in advisory committees established by the Regulatory Authority.
	Owner/User	• Be informed of purpose, use, and care of treatment system. • Be informed of existing rules and review and comment on any proposed program and/or rule changes. • Participate in advisory committees established by the Regulatory Authority.
	RME	• **Inform Owner/User of care and use of system.** • **Inform Owner/User of RME requirements and prohibited uses of system.**
PLANNING	Regulatory Authority	• Coordinate program rules and regulations with state, tribal, and local planning and zoning and other water-related programs. • Evaluate potential risks of wastewater discharges to limit environmental impacts on receiving environments during the rule making process. • Limit potential risks of environmental impacts from residuals management program and evaluate available handling/treatment capacities. • Inform local planning authority of rule changes and recommend their evaluation of potential impacts on land use.
	Developer	• Hire planners, certified site evaluators, and designers to ensure that all lots of proposed subdivision plats meet requirements for onsite treatment prior to final plat.
	RME	• **Develop criteria (e.g., site evaluation, design, construction) to be required of systems for acceptance into O&M program and inform Owners.** • **Continuously evaluate existing wastewater treatment needs and forecast future needs.**
PERFORMANCE	Regulatory Authority	• Establish system failure criteria to protect public health, e.g., wastewater backups in building, wastewater ponding on ground surface, insufficient separation from ground water or wells. • Establish minimum maintenance requirements for approved systems. • Establish performance criteria necessary to protect public health and water resources for each defined receiving environment in the Regulatory Authority's jurisdiction.
	Owner	• Regularly maintain system components in proper working order. • **Comply with any RME requirements regarding care and use of the system.**
	RME	• **Operate systems to comply with performance criteria stipulated in the operating permits.**
TRAINING AND CERTIFICATION/ LICENSING	Licensing Board/ Regulatory Authority	• Develop and administer training, testing, and certification/licensing program for site evaluators, designers, contractors, operators, pumpers/haulers, and inspectors. • Maintain a current certified/licensed Service Provider listing.
	Service Provider	• Obtain appropriate certification(s)/license(s) and continuing education as required. • Obtain training from the manufacturer or vendor regarding appropriate use, installation requirements, and operation and maintenance procedures of any proprietary equipment to be installed. • Comply with applicable federal, state, tribal, and local requirements in the evaluation of sites for wastewater treatment and dispersal.
	Owner	• When using third-party services, contract only with the appropriate certified/licensed Service Providers.
	RME	• **When using third-party services, contract with only the appropriate certified/licensed Service Providers.** • **Ensure that RME staff who operate and/or maintain systems obtain appropriate certification(s)/license(s) to practice.** • **Arrange for supplemental training as needed for Service Providers and/or staff to manage, operate, and/or maintain systems.**

[1] *Activities in bold are activities added to program elements from the preceding Management Model.*

MANAGEMENT MODEL 4: RME OPERATION AND MAINTENANCE

PROGRAM ELEMENT	RESPONSIBLE PARTY	ACTIVITY[1]
SITE EVALUATION	Regulatory Authority	• Codify prescriptive requirements for site evaluation procedures. • Codify criteria for treatment site characteristics suitable for permitted designs that will prevent unacceptable impacts on ground and surface water resources. • Establish the defining characteristics of each receiving environment in the Regulatory Authority's jurisdiction. • **Approve and oversee site evaluation procedures required by RME for system acceptance in the O&M program to ensure that system designs are appropriate for the sites and their stipulated performance criteria.**
	Site Evaluator	• Obtain certification/license to practice. • Describe site and soil characteristics, determine suitability of site with respect to code requirements, and estimate site's hydraulic and treatment capacity. • Comply with applicable federal, state, tribal, and local requirements in the evaluation of sites for wastewater treatment and dispersal.
	Owner	• Hire a certified/licensed site evaluator to perform site evaluation. • **Comply with any additional siting requirements established by RME for system acceptance in the O&M program.**
DESIGN	Regulatory Authority	• Codify prescriptive, pre-engineered designs that are suitable for treatment sites that meet the appropriate prescriptive site criteria. • Administer a plan review program for engineered designs to meet stipulated performance criteria. • Require submission of routine operation and emergency contingency plans that will sustain system performance and avoid unpermitted discharges.
	Designer	• Obtain certification/license to practice. • Design treatment system that is compatible with the site and soil characteristics described by the site evaluator. • Comply with applicable federal, state, tribal, and local requirements in the design of wastewater treatment and dispersal systems.
	Owner	• Hire a certified/licensed designer to prepare system design. • Comply with any additional design requirements established by the RME for system acceptance in the O&M program.
CONSTRUCTION	Regulatory Authority	• Administer a permitting program for system construction, including Regulatory Authority review of proposed system siting and design plans. • Require designer of record to certify that completed system construction is in substantial compliance with approved plans and specifications. • Require that record drawings of constructed system be submitted to the Regulatory Authority by Owner. • Require Owner to submit a copy of system O&M manual to the Regulatory Authority and RME.
	Contractor/ Installer	• Obtain certification/license to practice. • Construct system in accordance with the approved plans and specifications. • Prepare record drawings of completed system and submit to Owner. • Provide Owner with an O&M manual describing component manufacturer's maintenance and troubleshooting requirements/recommendations. • Comply with applicable federal, state, tribal, and local requirements in the design and construction of wastewater treatment and dispersal systems.
	Designer of Record	• Approve proposed field changes and submit to Owner. • Certify that construction of the system is substantially in conformance with the approved plans and specifications.
	Owner	• **Comply with any additional construction requirements established by the RME for system acceptance in the O&M program.** • Hire a certified/licensed designer to prepare system design. • Submit final record drawings of constructed system to Regulatory Authority. • Submit a copy of the system O&M manual to the Regulatory Authority and RME to record required maintenance.

[1] *Activities in bold are activities added to program elements from the preceding Management Model.*

MANAGEMENT MODEL 4: RME OPERATION AND MAINTENANCE

	PROGRAM ELEMENT	RESPONSIBLE PARTY	ACTIVITY[1]
MANAGEMENT MODEL 4: RME OPERATION AND MAINTENANCE	**OPERATION & MAINTENANCE**	Regulatory Authority	• Provide Owner/User with educational materials regarding system use and care. • Administer a program of renewable/revocable operating permits that are issued to RME, stipulating system performance criteria, compliance monitoring reporting schedule, term of permit, and renewal option upon documented compliance with operating permit stipulations. • Track and review compliance monitoring reports to ensure that systems are operating in accordance with operating permits. • **Consider replacing individual system operating permits with general permits issued to the RME for classes of systems.**
		Operator	• Inspect and service the system as necessary in accordance with the submitted O&M manual and/or operating permit stipulations. • Perform system monitoring as stipulated in RME's operating permit. • Certify to RME that the required maintenance and monitoring was performed in a timely manner and noting any system deficiencies. • Comply with applicable federal, state, tribal, and local requirements in the operation and maintenance of the treatment and dispersal system.
		Pumper/Hauler	• Obtain certification/license to practice. • Inspect and service system as necessary. • Comply with applicable federal, state, tribal, and local requirements in the operation and maintenance of treatment and dispersal system.
		Owner/User	• Follow recommendations provided by Regulatory Authority, Service Providers, and/or Owner to ensure that undesirable or prohibited materials are not discharged to system. • Maintain system components in proper working order. • **Comply with any RME requirements regarding care and use of system.**
		RME	• **Operate and maintain systems in accordance with the stipulated operating permit requirements.** • **Submit compliance monitoring reports to the Regulatory Authority according to the schedule stipulated in the operating permit.** • Hire a certified/licensed pumper/hauler or operator to maintain system.
	RESIDUALS MANAGEMENT	Regulatory Authority	• Administer a tracking system for residuals hauling, treatment, and disposal and review to evaluate compliance with 40 CFR Part 503 Use and Disposal of Sewage Sludge, 40 CFR Part 257, and applicable state, tribal, and local requirements. • Inventory available residuals handling/treatment capacities and develop contingency plans to ensure that sufficient capacities are always available.
		Pumper/Hauler	• Comply with applicable federal, state, tribal, and local requirements in the pumping, hauling, treatment, and disposal of wastewater treatment system residuals.
		RME	• Hire a certified/licensed pumper/hauler to remove, treat, and dispose of residuals. • **Comply with applicable federal, state, tribal, and local requirements in the pumping, hauling, treatment, and disposal of treatment system residuals.** • **Inventory available residuals handling/treatment capacities and develop contingency plans when insufficient capacities are available.**
	COMPLIANCE INSPECTIONS/ MONITORING	Regulatory Authority	• Perform inspection programs at point-of-sale, change-in-use of properties, "targeted areas," and/or systems reported to be in violation. • Conduct compliance inspections of residuals hauling, treatment, and disposal. • Administer a program to monitor timely submittals of acceptable compliance maintenance reports. • Perform system inspections randomly and/or at time of operating permit renewal.
		Inspector	• **Obtain certification/license to practice.** • **Perform system compliance inspections for RME in accordance with prevailing Regulatory Authority requirements.**
		RME	• **Submit compliance monitoring reports to the Regulatory Authority as stipulated in operating permit.** • **Submit compliance inspection report signed/sealed by a certified/licensed inspector prior to applying for renewal of operating permit.** • **Conduct regular reviews of management program with Owner/User and Regulatory Authority to optimize system operation program.** • **Hire a certified/licensed inspector to inspect system compliance status.**

[1] *Activities in bold are activities added to program elements from the preceding Management Model.*

PROGRAM ELEMENT	RESPONSIBLE PARTY	ACTIVITY[1]
CORRECTIVE ACTIONS	Regulatory Authority	• Negotiate compliance schedules with RME for correcting documented noncompliance items. • Administer enforcement program including fines and/or penalties for failure to comply with compliance requirements. • Obtain necessary authority to enter property to correct imminent threats to public health if the Owner/User fails to comply. • Require system inspection by certified inspector at time of operating permit renewal. • **Negotiate compliance schedules with RME, Owner/User, or both, for correcting documented noncompliance items.**
	Designer	• Provide Owner/RME with documents (drawings, specifications, modifications, etc.) that may be required by the Regulatory Authority prior to corrective actions.
	Contractor/ Installer	• Perform required repairs, modifications, and upgrades as necessary.
	Inspector	• Inspect treatment system for compliance with operating permit prior to permit renewal.
	Owner	• Comply with terms and conditions of the negotiated compliance schedule for component replacement/repairs. • Submit required documents for corrective actions to Regulatory Authority. • Hire appropriate certified/licensed Service Providers to perform required corrective actions.
	RME	• **Comply with terms and conditions of the negotiated compliance schedule for system performance.**
RECORD KEEPING, INVENTORY, & REPORTING	Regulatory Authority	• Administer a database inventory (locations, site evaluations, record drawings, permits, performed maintenance, and inspection reports) of all systems. • Maintain a residuals treatment and disposal tracking system. • Maintain a current certified/licensed Service Provider listing that is available to the public. • Administer a tracking system for operating permits. • Administer a tracking database for compliance reports. • **Administer periodic financial, management, and technical audits of RME.**
	Operator or Inspector	• Provide certified report of all maintenance and observed system deficiencies to RME. • Provide certified report of all observed system deficiencies to Owner. • Perform system monitoring as stipulated in RME's operating permit.
	Pumper/Hauler	• Prepare and submit records of residuals handling as required.
	Owner	• Maintain approved record drawings and O&M manual of system. • Maintain maintenance records of system. • Provide drawings, specifications, O&M manual, and maintenance records to new property owner at time of property transfer.
	RME	• **Maintain system monitoring and service records.** • **Inventory, collect, and provide permit information to Regulatory Authority.**
FINANCIAL ASSISTANCE & FUNDING	Regulatory Authority	• Provide the legal and financial support to sustain the management program. • Provide a listing of financial assistance programs available to Owner/User and the qualifying criteria for each program. • Consider implementing a state or local financing program to assist Owners in upgrading their systems.
	RME	• **Conduct regular reviews of management program with Owner/User and Regulatory Authority to optimize operations.**

[1] *Activities in bold are activities added to program elements from the preceding Management Model.*

MANAGEMENT MODEL 4: RME OPERATION AND MAINTENANCE

MANAGEMENT MODEL 5: RME OWNERSHIP

Objective: To provide professional management of the planning, siting, design, construction, operation, and maintenance of onsite/decentralized systems through Responsible Management Entities that own and manage individual and clustered systems within their service areas.

PROGRAM ELEMENT	RESPONSIBLE PARTY	ACTIVITY[1]
PUBLIC EDUCATION AND PARTICIPATION	Regulatory Authority	• Educate Owner/User on purpose, use, and care of treatment system. • Provide public review and comment periods of any proposed program and/or rule changes.
	Service Provider	• Be informed of existing rules, and review and comment on any proposed program or rule changes. • Participate in advisory committees established by the Regulatory Authority.
	RME	• Inform User of care and use of system. • Inform User of RME requirements and prohibited uses of system.
	User	• Be informed of purpose, use, and care of treatment system.
PLANNING	Regulatory Authority	• Coordinate program rules and regulations with state, tribal, and local planning and zoning and other water-related programs. • Evaluate potential risks of wastewater discharges to limit environmental impacts on receiving environments during the rule making process. • Limit potential risks of environmental impacts from residuals management program and evaluate available handling/treatment capacities. • Inform local planning authority of rule changes and recommend their evaluation of potential impacts on land use.
	Developer	• Hire planners, certified site evaluators, and designers to ensure that all lots of proposed subdivision plats meet requirements for onsite treatment prior to final plat.
	RME	• Continuously evaluate existing wastewater treatment needs and forecast future needs. • **Require developers to submit proposed subdivision plats to RME for review and comment to ensure compatibility with RME requirements.** • **Plan most cost-effective approach to meeting treatment needs through appropriate mix of central sewerage, clusters, and individual onsite systems.**
PERFORMANCE	Regulatory Authority	• Establish system failure criteria to protect public health, e.g., wastewater backups in building, wastewater ponding on ground surface, insufficient separation from ground water or wells. • Establish minimum maintenance requirements for approved systems. • Establish performance criteria necessary to protect public health and water resources for each defined receiving environment in the Regulatory Authority's jurisdiction.
	RME	• Operate, maintain, and repair systems to comply with performance criteria stipulated in the operating permits.
	User	• Comply with any RME requirements regarding care and use of the system.
TRAINING AND CERTIFICATION/ LICENSING	Licensing Board/ Regulatory Authority	• Develop and administer training, testing, and certification/licensing program for site evaluators, designers, contractors, pumpers/haulers, inspectors, and operators. • Maintain a current certified/licensed Service Provider listing.
	Service Provider	• Obtain appropriate certification(s)/license(s) and continuing education as required. • Obtain training from the manufacturer or vendor regarding appropriate use, installation requirements, and operation and maintenance procedures of any proprietary equipment to be installed. • Comply with applicable federal, state, tribal, and local requirements in the evaluation of sites for wastewater treatment and dispersal.
	RME	• When using-third party services, contract with only certified/licensed Service Providers. • RME staff who site, **design, construct,** operate, and/or maintain systems must obtain appropriate certification(s)/license(s) to practice. • Arrange for supplemental training as needed for Service Providers and/or staff to manage, operate, and/or maintain systems.

[1] *Activities in bold are activities added to program elements from the preceding Management Model.*

PROGRAM ELEMENT	RESPONSIBLE PARTY	ACTIVITY[1]
SITE EVALUATION	Regulatory Authority	• Codify prescriptive requirements for site evaluation procedures. • Codify criteria for treatment site characteristics suitable for permitted designs that will prevent unacceptable impacts on ground and surface water resources. • Establish the defining characteristics of each receiving environment in the Regulatory Authority's jurisdiction. • Approve and oversee site evaluation procedures used by RME to ensure that system designs are appropriate for the sites and their stipulated performance criteria.
	Site Evaluator	• Obtain certification/license to practice. • Describe site and soil characteristics, determine suitability of site with respect to code requirements, and estimate site's hydraulic and treatment capacity. • Comply with applicable federal, state, tribal, and local requirements in the evaluation of sites for wastewater treatment and dispersal.
	RME	• Hire a certified/licensed site evaluator to perform site evaluation.
DESIGN	Regulatory Authority	• Codify prescriptive, pre-engineered designs that are suitable for treatment sites that meet the appropriate prescriptive site criteria. • Administer the plan review program for engineered designs to meet stipulated performance criteria. • Require routine operation and emergency contingency plans that will sustain system performance and avoid the submission of unpermitted discharges.
	Designer	• Obtain certification/license to practice. • Design treatment system that is compatible with the site and soil characteristics described by the site evaluator. • Comply with applicable federal, state, tribal, and local requirements in the design of wastewater treatment and dispersal systems.
	RME	• Hire a certified/licensed designer to prepare system design.
CONSTRUCTION	Regulatory Design	• Administer a permitting program for system construction, including Regulatory Authority review of proposed system siting and design plans. • Require designer of record to certify that completed system construction is in substantial compliance with approved plans and specifications. • Require that record drawings of constructed system be submitted to the Regulatory Authority by RME.
	Contractor/ Installer	• Obtain certification/license to practice. • Construct system in accordance with the approved plans and specifications. • Prepare record drawings of completed system and submit to RME. • Provide RME with an O&M manual describing component manufacturer's maintenance and troubleshooting requirements/recommendations. • Comply with applicable federal, state, tribal, and local requirements in the design and construction of wastewater treatment and dispersal systems.
	Designer of Record	• Approve proposed field changes and submit to RME. • Certify that construction of the system is substantially in conformance with the approved plans and specifications.
	RME	• Hire a certified/licensed designer to prepare system design. • Submit final record drawings of constructed system to Regulatory Authority. • Submit a copy of system O&M manual to the Regulatory Authority to record required maintenance.

[1] *Activities in bold are activities added to program elements from the preceding Management Model.*

MANAGEMENT MODEL 5: RME OWNERSHIP

	PROGRAM ELEMENT	RESPONSIBLE PARTY	ACTIVITY[1]
MANAGEMENT MODEL 5: RME OWNERSHIP	OPERATION & MAINTENANCE	Regulatory Authority	• Provide User with educational materials regarding system use and care. • Administer a program of renewable/revocable operating permits that are issued to RME that stipulate system performance, compliance monitoring reporting schedule, term of permit, and renewal option upon documented compliance with operating permit stipulations. • Track and review compliance monitoring reports to ensure that systems are operating in accordance with operating permits. • Consider replacing individual system operating permits with general permits issued to RME for classes of systems.
		Operator	• Inspect and service system as necessary in accordance with the submitted O&M manual and/or operating permit stipulations. • Perform system monitoring as stipulated in RME's operating permit. • Certify to RME that the required maintenance and monitoring were performed in a timely manner and noting any system deficiencies. • Comply with applicable federal, state, tribal, and local requirements in the operation and maintenance of the treatment and dispersal system.
		Pumper/Hauler	• Obtain certification/license to practice. • Inspect and service system as necessary. • Comply with applicable federal, state, tribal, and local requirements in the operation and maintenance of the treatment and dispersal system.
		User	• Follow recommendations provided by Regulatory Authority, Service Providers, and/or Owner to ensure that undesirable or prohibited materials are not discharged to system. • Comply with any RME requirements regarding care and use of system.
		RME	• Operate and maintain systems in accordance with the stipulated operating permit requirements. • Submit compliance monitoring reports to the Regulatory Authority according to the schedule stipulated in the operating permit. • Hire a certified/licensed pumper/hauler or operator to maintain system.
	RESIDUALS MANAGEMENT	Regulatory Authority	• Administer a tracking system for residuals hauling, treatment, and disposal and review to evaluate compliance with 40 CFR Part 503 Use and Disposal of Sewage Sludge, 40 CFR Part 257, and applicable state, tribal, and local requirements. • Inventory available residuals handling/treatment capacities and develop contingency plans when capacities available are insufficient.
		Pumper/Hauler	• Comply with applicable federal, state, tribal, and local requirements in the pumping, hauling, treatment, and disposal of wastewater treatment system residuals.
		RME	• Hire a certified/licensed pumper/hauler to remove, treat, and dispose of residuals. • Comply with applicable federal, state, tribal, and local requirements in the pumping, hauling, treatment, and disposal of treatment system residuals. • Inventory available residuals handling/treatment capacities and develop contingency plans when capacities available are insufficient.
	COMPLIANCE INSPECTIONS/ MONITORING	Regulatory Authority	• Perform inspection programs at point-of-sale, change-in-use of properties, "targeted areas," and/or systems reported to be in violation. • Conduct compliance inspections of residuals hauling, treatment, and disposal. • Administer a program to monitor timely submittals of acceptable compliance maintenance reports. • Perform system inspections randomly and/or at the time of operating permit renewal.
		Inspector	• Obtain certification/license to practice. • Perform system compliance inspections for RME in accordance with prevailing Regulatory Authority requirements.
		RME	• Submit compliance monitoring reports to Regulatory Authority as stipulated in operating permit. • Submit a compliance inspection report signed/sealed by a certified/licensed inspector prior to applying for renewal of operating permit. • Conduct regular reviews of management program with Regulatory Authority to optimize system operation program. • Hire a certified/licensed inspector to inspect system compliance status.

[1] *Activities in bold are activities added to program elements from the preceding Management Model.*

PROGRAM ELEMENT	RESPONSIBLE PARTY	ACTIVITY[1]
CORRECTIVE ACTIONS	Regulatory Authority	• Negotiate compliance schedules with RME for correcting documented noncompliance items. • Administer the enforcement program including fines and/or penalties for failure to comply with compliance requirements. • Require system inspection by a certified inspector at time of operating permit renewal. • **Negotiate compliance schedules with RME for correcting documented noncompliance items.**
	Designer	• Provide RME with documents (drawings, specifications, modifications, etc.) that may be required by the Regulatory Authority prior to corrective action.
	Contractor/ Installer	• Perform required repairs, modifications, and upgrades as necessary.
	Inspector	• Inspect treatment system for compliance with operating permit prior to permit renewal.
	RME	• **Comply with terms and conditions of the negotiated compliance schedule.** • **Submit required documents for corrective actions to the Regulatory Authority.** • **Hire appropriate certified/licensed Service Providers to perform required corrective actions.**
RECORD KEEPING, INVENTORY, & REPORTING	Regulatory Authority	• Administer a database inventory (locations, site evaluations, record drawings, permits, and inspection reports) of all systems within the Regulatory Authority's jurisdiction. • Maintain a residuals treatment and disposal tracking system. • Maintain a current certified/licensed Service Provider listing, which is available to the RMEs. • Administer a tracking system for operating permits. • Administer a tracking database for compliance reports. • Administer financial, management, and technical audits of RME.
	Operator or Inspector	• Provide a certified report of all maintenance and observed system deficiencies to RME. • Provide a certified report of all observed system deficiencies to Owner. • **Perform system monitoring as stipulated in RME's operating permit.**
	Pumper/Hauler	• Prepare and submit records of residuals handling as required.
	RME	• **Maintain system monitoring and service records.** • **Inventory, collect, and provide permit information to Regulatory Authority.**
FINANCIAL ASSISTANCE & FUNDING	Regulatory Authority	• Provide the legal and financial support to sustain the regulatory program. • Provide a listing of financial assistance programs available to RME and the qualifying criteria for each program. • Consider implementing a state or local financing program to assist RME in upgrading systems.
	RME	• **Conduct regular reviews of management program with Regulatory Authority to optimize operations.**

MANAGEMENT MODEL 5: RME OWNERSHIP

[1] *Activities in bold are activities added to program elements from the preceding Management Model.*

APPENDIX B:
RELATIONSHIP TO OTHER EPA WATER PROGRAMS

APPENDIX B: RELATIONSHIP TO OTHER EPA WATER PROGRAMS

The Management Guidelines will help support the activities and approaches being applied in several other EPA programs and contribute toward achieving mutual water quality objectives and public health protection goals. Related programs include watershed management, water quality management, biosolids and residuals management, nonpoint source control, source water assessment and protection, underground injection control, water permitting, and coastal zone management. The relationship of the Management Guidelines to these companion programs is summarized in the following discussion.

Watershed Management. The Management Guidelines can be integrated into a comprehensive watershed approach at the state, tribal, or local government level. There are clear benefits to managing onsite/centralized systems at the basin, watershed, or subwatershed level. Ideally, the use of a watershed approach will facilitate the identification of both existing and anticipated sources of pollutants of concern, e.g., nutrient and pathogens, and allow the appropriate jurisdictions to take coordinated actions to protect or restore an identified resource. In such an approach, short- and long-term wastewater management plans and actions for both centralized and decentralized systems can be integrated into a comprehensive plan that may include analyses and actions that address the impacts of other contributing sources of pollutants such as animal waste, wildlife, or agriculture. The use of a watershed approach also encourages the coordination of management entities and actions across jurisdictions. Interjurisdictional planning and coordination can result in more efficient resource utilization, including data sharing, and also help to avoid inconsistent management policies or requirements that can cause unanticipated

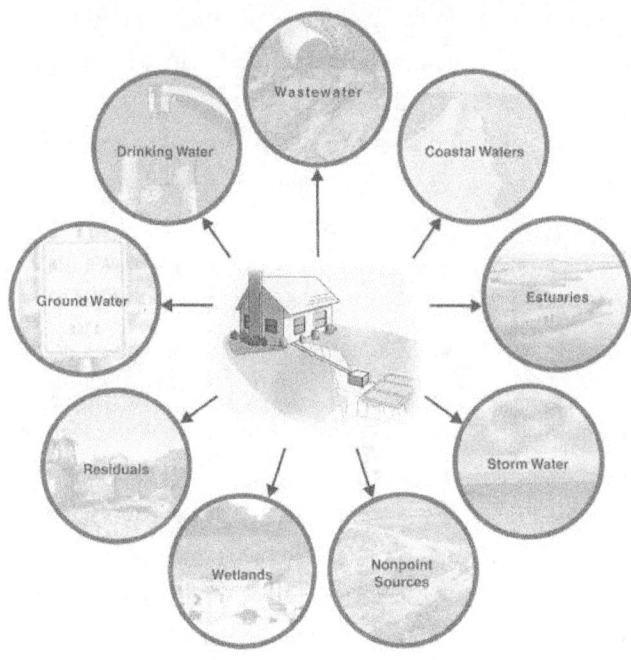

consequences, such as accelerated growth in adjacent communities due to their less burdensome requirements or lower costs.

National Pollutant Discharge Elimination System (NPDES). In 1972 Congress established the NPDES program under the Clean Water Act (CWA). Under the CWA, discharge of a pollutant from a point source to waters of the United States is prohibited unless that discharge is authorized by an NPDES (CWA Section 402) or wetlands (CWA Section 404) permit. The NPDES program includes discharges to ground water with a direct hydrologic connection to surface water. NPDES permits are issued by a state or tribe authorized to implement the NPDES program, or by EPA if there is no authorized state or tribe. The NPDES permit establishes necessary technology-based and water quality-based terms, limitations, and conditions on the discharge to protect public health and the environment.

EPA's NPDES regulations (40 CFR 122.28) provide for issuance of a "general permit" to authorize discharges from similarly situated facilities such as onsite and clustered systems. Several states, including Arkansas, Kentucky, and North Carolina, issue general permits. The draft Management Handbook contains an example of the key aspects of a general permit.

Biosolids and Residuals Management. The 1987 Amendments to the CWA required the development of comprehensive requirements for the use and disposal of sewage sludge (biosolids). As defined in the resulting "Use and Disposal of Sewage Sludge" rule at 40 CFR Part 503, *sewage sludge* includes the residuals produced by the treatment of domestic sewage (other than grit and screenings) and includes septage from onsite and clustered wastewater treatment systems. The Part 503 rule (along with the nonhazardous solid waste disposal requirements under 40 CFR Parts 257 and 258, which apply when domestic septage is mixed with other waste sources by pumpers) establish minimum federal requirements for the proper management of septage from onsite and clustered wastewater treatment systems. EPA has developed supplemental guidance on the management of septage in *Domestic Septage Regulatory Guidance: A Guide to the EPA 503 Rule*[13] and *Guide to Septage Treatment and Disposal*[14].The use and disposal of sewage sludge is usually regulated as part of the NPDES program.

Storm Water Management. Historically, polluted storm water runoff was often transported by municipal separate storm sewer systems (MS4s) or discharged from industrial or construction activities and ultimately discharged into local rivers and streams without treatment. Common pollutants include oil and grease from roadways, pesticides from lawns, sediment from construction sites, and carelessly discarded trash, such as cigarette butts, paper wrappers, and plastic bottles. When deposited into nearby waterways through MS4 discharges, these pollutants can impair the waterways, thereby discouraging recreational use of the resource, contaminating drinking water supplies, and interfering with the habitat for fish, other aquatic organisms, and wildlife.

In 1990 EPA promulgated rules establishing Phase I of the National Pollutant Discharge Elimination System (NPDES) storm water program. The Phase I program requires communities with MS4s serving populations of 100,000 or greater or sites with industrial or construction activity to implement a storm water management program as a means to control polluted discharges. The Storm Water Phase II Rule, promulgated on December 8, 1999, extends coverage of the NPDES storm water program to certain "small" MS4s and small construction sites. Operators of regulated small MS4s are required to design their programs to reduce the discharge of pollutants to the "maximum extent practicable," protect water quality, and satisfy the appropriate water quality requirements of the Clean Water Act.

The Phase II program for MS4s is designed to accommodate a general permit approach using a Notice of Intent (NOI) as the permit application. The operator of a regulated small MS4 must include in the permit application, or NOI, its chosen best management practices (BMPs) and measurable goals for each of six minimum control measures. To help permittees identify the most appropriate BMPs for their programs, EPA will issue a "menu" of BMPs to serve as guidance.

One measure in a Phase II storm water program is the detection and elimination of illicit discharges. EPA has determined that many onsite and clustered systems (typically those that discharge to surface waters) illicitly discharge effluent to storm ditches that drain to storm sewers. In such cases there must be a permit approach to protect the MS4 from pollutants associated with the onsite and clustered system. The Management Guidelines can be used to assist NPDES permit applicants in determining appropriate BMPs.

Water Quality Management (including Total Maximum Daily Loads). Nationally, states have reported in their Clean Water Act Section 303(d) reports that designated uses are not being met for approximately 5,400 water bodies because of pathogens and that approximately 4,700 water bodies are impaired by nutrients[12]. Onsite wastewater treatment systems are often significant contributors of pathogens and nutrients. Under EPA's current requirements a Total Maximum Daily Load (TMDL) determination is required when the total loading of pollutants to a water body results in a violation of water quality standards. The Agency promotes the control and management of both point and nonpoint source discharges on a watershed basis. If onsite and clustered systems are determined to be a significant source of the pollutants, increased management is needed.

The most common approach to resolving problems with onsite wastewater treatment systems has been to replace the systems with a centralized wastewater treatment and collection system. However, a decentralized approach, with a high level of management, is capable of meeting water quality objectives while offering communities a wider range of options. In these situations, the Management Guidelines can be a valuable tool to use as the basis of TMDL/watershed implementation plans that promote improved management to address identified problems. An appropriate level of management, as described in this document, could reduce pollutant loads to achieve water quality standards. EPA also recognizes, as discussed more fully above, that there are situations where a system is subject to the NPDES program. In such cases, permit requirements should be consistent with any applicable TMDL and water quality standards.

Water Quality Standards. State and tribal water quality standards do not consistently address pathogen and nutrient loadings. This lack of consistency has resulted from a scarcity of information on how to measure, monitor, and evaluate the impacts of pathogens and nutrients on water quality. New methods and information are being developed to assist tribes, states,

and local governments in assessing and developing appropriate management strategies to control these pollutants. EPA is developing recommendations for improved methods to measure and document human health risks due to exposure to the most common pathogens and differing concentrations of these pathogens. A thorough discussion is available in the draft *Implementation Guidance for Ambient Water Quality Criteria for Bacteria-1986.*[15] EPA is also developing a series of *Nutrient Criteria Technical Guidance Manuals*[16][17][18] for various water body types, such as rivers and streams. The intent of these documents is to provide states and tribes with methods to assess waterbody nutrient impairment, select criteria, design monitoring programs, and implement management practices. These factors should be considered during the siting, design, and operation of onsite and decentralized wastewater treatment systems.

Source Water Assessment and Protection. The 1996 Amendments to the Safe Drinking Water Act (SDWA) require states and tribes to implement Source Water Assessment and Protection (SWAP) programs that assess areas serving as sources of drinking water, identify potential threats, and implement protection efforts. The SWAP requires states to conduct source water assessments for all their public water systems. Assessments consist of delineating protection areas for the source waters of public drinking water supplies, identifying potential sources of contaminants within these areas, determining the susceptibility of the water supplies to contamination from these potential sources, and making the results of the assessments available to the public. Assessments for many water systems, such as those in rural areas, are likely to inventory onsite and clustered systems located in delineated source water protection areas and identify some of them as priority pollution threats. Communities are encouraged to consider this emerging information from the assessments as a factor in deciding what level of management of onsite and clustered systems is necessary. Several programs specifically address the protection of ground water because it serves as the source of drinking water for 95 percent of the nation's population in rural areas and for half of the total U.S. population. EPA also recommends the Management Guidelines as a tool in the protection of drinking water sources.

Underground Injection Control (UIC) Program. Certain onsite systems are regulated under the UIC program. The UIC program was established by the SDWA to protect current and future underground sources of drinking water (USDWs) from contamination caused by subsurface disposal of wastes. EPA groups underground injection into five classes (Classes I–V), from deep to shallow. Class V wells include typically shallow, percolating systems, such as dry wells, leach fields, and similar types of drainage wells that overlie USDWs.

Under the existing federal regulations, most Class V injection wells are authorized by rule provided they meet certain reporting requirements (e.g., submit inventory information) and do not endanger USDWs. EPA recognizes that state, tribal, and local governments commonly regulate onsite systems of varying sizes. Regardless, the UIC program is responsible for ensuring that these entities meet UIC program requirements when regulating large-capacity septic systems (those that accept solely sanitary waste and have the capacity to serve 20 or more people per day). Onsite wastewater treatment systems may also be regulated under the UIC program by an authorized state, tribe, or EPA if they accept industrial, chemical, or other non-sanitary wastes, also called "industrial drainage wells" or "agricultural drainage wells."

In 1999 the UIC program undertook two efforts relevant to large-capacity septic systems. First, the program promulgated regulations prohibiting the construction of new large-capacity cesspools and ordered all existing large-capacity cesspools to be closed by April 5, 2005. Second, the program completed a comprehensive study of shallow injection wells, including septic systems, that are regulated under the UIC program.[19] EPA found that although the prevalence of contamination cases appears low relative to the prevalence of these systems, there are documented examples that implicate these large systems as sources of ground water contamination, and they are being addressed locally.

On June 7, 2002 (67 FR 39583), EPA announced a final determination for all subclasses of Class V wells (such as large-capacity septic systems) not included in the December 7, 1999, final UIC rule. The Agency determined that additional federal requirements are not needed at this time and that existing federal underground injection control regulations are adequate to prevent Class V wells from endangering USDWs. This determination is based on the actions EPA is taking to improve the performance of onsite and clustered systems through the development of the Management Guidelines.

Coastal Zone Managment Act/Coastal Zone Act Reauthorization Amendments of 1990 (CZMA/CZARA). EPA and the National Oceanic and Atmospheric Administration (NOAA) jointly administer Section 6217 of the CZMA/CZARA. This provision requires the 29 states with approved Coastal Zone Management Programs to establish and implement Coastal Nonpoint Pollution Control Programs. These programs must include management measures for both new and operating onsite sewage dispersal systems (OSDS). The measures are described in EPA's *Guidance Specifying Management Measures for Sources of Nonpoint Pollution in Coastal Waters*[20]. The measure for new OSDS specifies that they be designed, installed, and operated properly and be situated at safe distances from sensitive resources, including wetlands and floodplains. Protective separation between the bottom of the infiltration system and ground water tables is to be established, and OSDS are to be designed to reduce nitrogen loadings in areas where surface waters might be adversely affected. The measure for operating OSDS requires operation and maintenance to prevent surface water discharge and reduce loadings to ground water, as well as inspection at regular time intervals and repair or replacement of faulty systems. The OSDS measures described above are consistent with many of the concepts described in the Management Guidelines.

Nonpoint Source Program. Congress established the national nonpoint source program in 1987 when it amended the Clean Water Act with Section 319. States were required to conduct nonpoint source assessments and develop EPA-approved "Nonpoint Source Management Programs." All states and territories and, as of September 2001, more than 70 tribes (representing over 70 percent of Indian lands) now have EPA-approved nonpoint source assessments and management programs. Typical categories of nonpoint sources identified and addressed in the state, territorial, and tribal assessments and management plans include agriculture, urban, onsite disposal systems, forestry, and hydromodification. In some states the primary responsibility for managing onsite and clustered systems falls within the purview of the nonpoint source program.

Congress provides funding to assist the states, territories, and tribes in developing and implementing their nonpoint source management programs. These funds can be used by states, territories, and tribes to address sources identified in their management program submissions. States, territories, and tribes can use these funds to promote, demonstrate, and fund activities relating to onsite and clustered management programs, including monitoring, program assessments and development, demonstration projects, research, public education and outreach, and system replacement or rehabilitation. The voluntary Management Guidelines are intended to support the achievement of the goals of the state, territorial, and tribal programs as they relate to onsite and clustered program management.

Technology Transfer. EPA recently published the *Onsite Wastewater Treatment Systems Manual*[21] (Onsite Manual) to provide new information on alternative treatment technologies and to promote a performance-based approach to onsite and clustered wastewater system management. This document is an update of EPA's 1980 *Design Manual - Onsite Wastewater Treatment and Disposal Systems*[22]. The Onsite Manual serves as the technical complement to the Management Guidelines and as a reference to identify the environmental, technological, administrative, and public health factors to consider when developing an improved management program. The Onsite Manual contains information that program managers can use in assessing the environmental impacts of specific onsite and clustered wastewater treatment technologies on both the watershed and individual site levels and in the selection of appropriate technologies.

NOTES

NOTES

www.ingramcontent.com/pod-product-compliance
Lightning Source LLC
Chambersburg PA
CBHW080647180526
45168CB00008B/3334